Global Climate Justice

Proposals, Arguments and Justification

Olivier Godard

Senior Research Associate, Department of Economics, École Polytechnique, France

Edward Elgar
PUBLISHING

Cheltenham, UK • Northampton, MA, USA

Published by
Edward Elgar Publishing Limited
The Lypiatts
15 Lansdown Road
Cheltenham
Glos GL50 2JA
UK

Edward Elgar Publishing, Inc.
William Pratt House
9 Dewey Court
Northampton
Massachusetts 01060
USA

A catalogue record for this book
is available from the British Library

Library of Congress Control Number: 2017939807

This book is available electronically in the **Elgar**online
Social and Political Science subject collection
DOI 10.4337/9781786438157

MIX
Paper from
responsible sources
FSC
www.fsc.org
FSC® C013056

ISBN 978 1 78643 814 0 (cased)
ISBN 978 1 78643 815 7 (eBook)

Typeset by Servis Filmsetting Ltd, Stockport, Cheshire
Printed and bound in Great Britain by TJ International Ltd, Padstow

Contents

Boxes

Preface

The analyses and reflections presented in this book have been nourished by more than 25 years of observation and thinking concerning policy responses to climate change at the national and international levels. The work began in 1990 at the International Centre for Research on Environment and Development (Centre international de recherche sur l'environnement et le développement, CIRED), a joint unit of two French research institutions, the École des Hautes Études en Sciences Sociales and the French National Centre for Scientific Research (Centre national de la recherche scientifique, CNRS), under the dynamic stimulus of Jean-Charles Hourcade, a key contributor to research on climate and energy economics on the French intellectual scene over all those years. When we began, one of the key debates among economists concerned the type of instruments that would best fit the climate issue: carbon taxes or tradable permits? Another was about the pace at which the world economy should switch to a low-carbon profile, taking account of pervasive uncertainty, prospects for learning and the limits of adaptation capacities in case of climate shock. A third key debate focused on what would be the best strategy in terms of building an international coalition for action. Over the years I observed how, for legal and political reasons, carbon taxes were dismissed in various countries, notably in France, as well as at the international level, in favour of quantitative emissions targets and, subsequently, of emissions trading: the first choice in favour of quantitative limits on greenhouse gas emissions and emissions trading was made in Rio de Janeiro in 1992 with the adoption of the United Nations Framework Convention on Climate Change (UNFCCC), which was later confirmed and amplified in 1997 in Kyoto. The choice of emissions trading has been pushed back to the margins, but not wholly dismissed, in the Copenhagen and Paris Agreements thereafter.

In fact, more than national taxes, international cap-and-trade schemes require an explicit agreement to be reached on crucial distributive issues: how should we allocate emissions quotas or binding emissions abatement targets to countries? In the early 1990s, I saw this requirement as asking too much of international coordination to be a realistic approach. This was quite a different view from the United States (US) proposal being pushed at the time by Richard Stewart and Jonathan Wiener (see their 2003 book) of the US Department of Justice in favour of a 'comprehensive approach' based on national quantitative limits for all countries and all gases, supplemented by a world market of carbon quotas. To reach my own conclusion, I used the recent works on justification that were then being developed in France under the impetus of Luc Boltanski and Laurent Thévenot. I am grateful to these two social scientists for providing intellectual resources that enabled me to make a leap in understanding the specifics of conflicts concerning environmental issues.

When I joined the Laboratoire d'Econométrie at the École Polytechnique in Paris in 1998, my first task was to prepare a report on emissions trading in the context of climate change policies for the Council of Economic Analysis, reporting to the French Prime Minister Lionel Jospin (Godard and Henry, 1998). This request came from Claude Henry, with whom I cooperated for the report and to whom I am indebted for having invited me to join the Laboratoire d'Econométrie, the team that he formed in the 1970s at the École Polytechnique. Claude Henry was then well known in the field of public economics. He was the author of the seminal paper published in the *American Economic Review* in 1974 on the 'irreversibility effect' under uncertainty.

The next decade, the 2000s, saw increasing international tensions about the enforcement of the Kyoto Protocol and the design of a post-Kyoto regime. A growing contradiction was observed between the framework set by the UNFCCC and its subsequent protocol, with a huge divide between the industrialized countries and the so-called developing countries, and the actual evolution of greenhouse gas emissions and economic power in the different regions of the world. At the same time, some rather extravagant statements and proposals were made under the umbrella of climate justice. I then stepped up my work on this issue, combining a critical examination of arguments in moral philosophy with a mobilization of the conceptual framework provided by the theory of justification, which belongs to the realm of moral sociology and offers, I believe, some penetrating and original insights into these issues.

The work presented in this book has benefited from research on climate change economics developed at the Laboratoire d'Econométrie with the cooperation of Jean-Pierre Ponssard, and the support of the Électricité de France (EDF) Chair in Sustainable Development at École Polytechnique. It has also been stimulated by the Climate Governance Programme of the European University Institute (EUI) in Florence, further to invitations in 2011 and 2012 from Denny Ellerman, a part-time Professor at this Institute. I would like to thank Jean-Pierre and Denny for their confidence and our intellectual interactions.

I acknowledge my intellectual debt to the colleagues I have named and to many others whom I cannot cite, but I remain solely responsible for the analyses I put forward on climate justice, justification and international cooperation. Should a reader disagree with my analyses, they must blame me rather than justification theory or the research institutions to which I am affiliated. I cannot end this Preface without addressing my heartfelt thanks to Anna Kiff, who has undertaken the difficult task of English editing for this book.

NOTE ON THE TEXT

This book has been developed from several works published since the late 1980s. The main recent resource for chapters 4, 6 and 7 is to be found in a book I released in French in October 2015: *La justice climatique mondiale*, published in Paris by Éditions La Découverte (Series Repères 651). I am grateful to Hugues Jallon, the CEO of Éditions La Découverte, for granting me the permission to use this material.

Apart from a few ad hoc additions reflecting recent events, the text was finalized in the fall of 2016.

Let me address a final note to the reader on how they should understand references to the literature given throughout the text: when it is not explicit, a reference to an author about an argument or an opinion does not mean that the author defends this argument or position, but only that he proposes a reflection on it – whether critical or supportive.

1. Introduction

The threat of anthropogenic climate change is caused by worldwide human emissions of greenhouse gases (GHGs). It is a threat of global scale. All the components of the Earth, its terrestrial ecosystems, oceans, atmosphere and their relationships will be impacted upon and transformed, maybe for millennia (Archer, 2008). We are facing what the American philosopher Stephen Gardiner (2011) called 'a perfect moral storm'. On a geological timescale, all dimensions of human existence and activity will be affected. It is in the common interest of humankind to limit the scale and pace of these global changes to the conditions of life on Earth. The need for convergent international action is brought to its highest level: almost all people, through the states that represent them, but also by involving the various groups and organizations, from cities and regions to non-governmental organizations (NGOs) and businesses, must converge on the ways and means of ensuring the effective prevention of a possible catastrophic scenario. Time is now very limited for meeting these requirements due to the inertia of the physical processes involved and the delay imposed until 2015 on any serious action towards low-carbon technologies and economies, despite the adoption of the United Nations Framework Convention on Climate Change (UNFCCC) in 1992.

To ensure that all people converge on expectations and priorities and commit to effective collective action, they must find a mutually acceptable answer to the issues of justice raised by the climate threat, but also by the existing state of the global economic (dis)order. For the common good of climate protection, will everyone, every community and every country, be fairly charged? On the international stage, all parties involved have expressed their desire to find 'fair and equitable' solutions to all practical aspects of the problem: the level of ambition for controlling GHG emissions; the distribution of efforts and costs; the conditions for adaptation to the already inevitable signs of climate change; the financial and technological transfers to be organized in support of climate-friendly actions and relief for the poorest. But serious disagreements arise when it comes to giving both theoretical foundations and practical meaning to these commitments to fairness and equity.

Those in the developing world who consider themselves to be essentially the present and future victims of climate change ask those in the developed countries, which they consider to be the originators of the global threat, to assume a 'historical responsibility'.[1] By this, they mean that the developed countries should pay compensation and receive a far smaller share of the carbon budget still available for the future. At the same time, they have asserted their own right to unconstrained development, and initially saw calls for common action in favour of climate protection as highly dubious, since the developed countries were not behaving according to their expectations. The organization of climate negotiations has been regularly criticized for failing to give each party, and in particular the less-developed countries and local communities, fair representation and access to appropriate expertise. The developed countries have said that they are ready to consider the specificities of the developing countries, and to help them. But they have refused to acknowledge any obligation to pay compensation or to be penalized in the name of their historical emissions.

Some basic questions are raised: Who can legitimately be said to be responsible for what? How can the international community ensure actions aimed at controlling GHG emissions do not lock hundreds of millions of people into poverty and deprive them of minimum access to energy, which is essential to cover the basic needs of all (Shue, 2014; Moellendorf, 2014)? How should we tackle the huge differences in resources, income and lifestyles across the planet? Is it possible to simultaneously drive a major programme aimed at eradicating poverty and engage a rapid worldwide shift away from carbon energy sources, as is assumed by the 17 Sustainable Development Goals adopted by the United Nations in September 2015? It is impossible to think about such issues without referring, whether explicitly or implicitly, to general views on the requirements of justice and equity in an international context.

The concern for justice marked the first principles laid down in the UNFCCC text. Its article 3.1 reads:

> The Parties should protect the climate system for the benefit of present and future generations of humankind, on the basis of equity and in accordance with their common but differentiated responsibilities and respective capabilities. Accordingly, the developed country Parties should take the lead in combating climate change and the adverse effects thereof.

Despite this agreement on the three principles of 'equity', 'common but differentiated responsibility' and 'capabilities', the international community failed to go very far in finding a precise common understanding of the practical means to give them substance. In 1992, the Parties had agreed that the developed countries – the Organisation for Economic Co-operation and Development (OECD), Eastern European and former

Soviet Union countries – should take the lead in the fight against climate change, and provide financial support to the developing countries for 'agreed' actions that the latter would undertake specifically for the protection of the global climate. This resulted in a division of the world into two groups, easily identified as the North and the South, to which differentiated obligations were assigned. The Kyoto Protocol, which was adopted in 1997 but only entered force in 2005, defined quantitative targets for controlling GHG emissions exclusively by the Northern Parties. Targets were also differentiated among these countries. The USA has never ratified this protocol, and Canada withdrew from it when it was enforced. The 2000s failed to show any significant progress in the organization of international action. The hopes placed in the December 2009 Copenhagen Conference of the Parties (COP 15) of coming to a decisive legally binding agreement were disappointed, then transferred to the Paris Conference of the Parties of December 2015 (COP 21).

The latter was considered a success because it arrived at the 'Paris Agreement' that preserves the existence of a full international cooperation framework on the climate issue, although it was clouded by the announcement of US withdrawal by President Trump in June 2017. In any event, it remained marked by a major contradiction between the objective announced (limiting temperature rise to significantly below 2°C, with the hope of remaining closer to 1.5°C) and the national objectives and plans disclosed at the time by the Parties which, taken together, have shown a global emissions control rate that was half of the figure needed to reach the global objective.

Over the years since 1992, global GHG emissions have grown at a sustained pace, despite reductions in European countries, and atmospheric concentrations of GHGs have inevitably followed suit, reaching levels higher than any seen in the last half-million years. Governments have lacked the motivation to do what moralists tell them they should do. But is it so surprising that governments are reluctant to impose profound changes and huge economic and political costs on their populations and constituencies for a benefit that will not be perceived for several decades, with its larger part benefiting future populations in other countries? It is often said that designing a fair and just agreement is critical for its general acceptance and its ultimate success at the implementation stage; efficiency would imply seeking justice. However, the various claims put forward in the name of climate justice, often extravagant and ill-founded, have contributed to turn the requirements of climate justice into an additional obstacle to reaching a sound agreement between the parties (Pickering and Barry, 2012).

The context is therefore a very difficult one. An agreement on some abstract principles is far from delivering a full agreement on what it means in practice or resolving the motivational factor. At the very least, an intellectual clarification is needed to test the viability and usefulness of ideas of justice in enhancing the broad convergence of national economies towards climate-friendly, low-carbon technologies in more than 30 years' time; in other words, the 2050 time horizon.

The purpose of this book is to present and discuss the diversified and conflicting intellectual landscape of the idea of global climate justice. The latter can be driven towards either cosmopolitanist moral approaches or a more politically oriented international framework. It is an idea that is highly exposed to the risk of being reduced to naive moral calls, simple ideological slogans or political gesticulations from stakeholders and parties to negotiations.

The landscape of this idea has been built via two different approaches, which have had evident links both ways. Since 1990, international climate negotiations have been accompanied by a political process involving the conceptualization of a new ideological corpus based on the denunciation of inequalities and injustices imposed on the developing countries, or the most vulnerable populations of these countries, due to global climate change. Concepts of the 'historical responsibility' and 'ecological and climate debt' of developed countries have been put forward. These have been used by activist scientists, NGOs and some governments to enhance and disseminate the conviction that the North was the only culprit and should pay on different lines: huge compensation for existing and expected damage and for additional obstacles to development in the South; free technology transfer and assistance to facilitate adaptation and enhance development; and allocation of the smallest share, if any, of future carbon budgets, because of its past overconsumption of common natural resources since the beginning of the industrial revolution in 1850. At the same time, equity has been asserted in order to demand equal per capita allocation of GHG emissions rights to countries, the atmosphere being said to be the common property of humankind in which every human being has the same right. Overall, this ideological corpus has shaped a denunciatory and adversarial approach to climate justice, articulated with the assertion of new rights concerning environmental quality and prospects for economic development.

Both in parallel and in interaction with this political process, scholars have developed concepts and arguments to support or criticize views on global and international climate justice and their palatability for climate negotiations. Moral thinking, political philosophy and economics provided the main frameworks to drive this process. Key themes developed concern the extension of domestic norms of justice to the international level; the content of a cosmopolitanist approach to justice; intergenerational justice and future ethics; moral and legal human rights and duties; and the pros and cons of the 'historic responsibility' concept.

The presentation of this dual landscape surrounding the same key concepts is intended to inform the critical examination of proposals made for the negotiation process as well as theoretical arguments put forward by scholars to support their views on the norms of justice and the rights and duties of different agents. Such a discussion cannot ignore the more general debates raised by theories of justice, but I will focus my account of discussions on those applications of theories that are explicitly directed at international relationships and climate change. In some cases, there are good reasons behind each of the main conflicting positions, but in other cases there are bad reasons or unconvincing arguments behind the proclaimed factual or moral evidence.

In this book, two aspects of the discussion will be particularly questioned:

1. Extending to the global level normative propositions hitherto designed for a 'well-ordered society' framed as a sovereign democratic nation state, one that has a territory marked by borders. This classical framework enabled the distinction between two types of norms: one for the internal justice of the society considered, and the other for organizing relations with people living in other countries. Is short-cutting this dual scheme a path to follow? What should we think of a cosmopolitanist viewpoint that considers the division of humankind into 200 nation states as an arbitrary heritage of history and an obstacle to the full acknowledgement of universal rights and values?
2. Bringing the natural environment of human activities into the intellectual circle of justice, based on the rights and obligations of human agents. Which concepts are the most appropriate to capture empirical expressions of climate change and weather events? Should we acknowledge a natural right to a stable climate? Can we legitimately base solutions for the twenty-first century on a retroactive application of the rights and obligations that some authors support had been those of the states from the beginning of the industrial revolution?

As a matter of fact, for 25 years, climate negotiations have been looking for solutions that could be accepted by all parties as fair, equitable and efficient, but they have not actually found them, even if the Chinese representative was authorized to declare, when the time came to approve the Paris Agreement (12 December 2015), that 'the Paris Agreement is fair and just, comprehensive and balanced, highly ambitious, enduring and effective, and with legally binding force' (Dimitrov, 2016).

Chapter 2 of this book presents an overview of concepts and distinctions about theories of justice and introduces two focal theories: that of John Rawls, and utilitarianism. Chapter 3 explores the relationships between justice and international coordination. It first examines how justice considerations may have a different role in international relations depending on the kind of model used to approach the international world. Next, it looks at how international relations may come into theories of justice, giving the spectrum of positions.

Chapter 4 introduces issues of justice raised by climate change. Elements of institutional history and objective data are given, before the main proposals made by various sources (NGOs, experts, governments and scholars) are presented.

Chapter 5 considers the issue of intergenerational equity, first by introducing the debate on standard economic discounting and then by discussing key ethical points. The issue of freedom of choice acknowledged to future generations is at the heart of the discussion of duties to the dead and the unborn, for which Janna Thompson (2009) proposed a challenging foundation. The classical but unresolved problems of non-existence and non-identity are introduced and discussed. All this concludes with the fragility of the idea of intergenerational justice between distant generations.

Chapter 6 provides a critical examination of the thesis of the 'historic responsibility' of countries for their emissions between 1850 and 1990. In fact, in argument after argument, it shows why we may doubt that this thesis is well founded either empirically or normatively.

Chapter 7 is future-oriented and considers the possible criteria for an equitable sharing of the carbon budget between countries. It begins with the issue of rights, which is not mentioned by the UNFCCC, and then continues with the three principles put forward by this international agreement: equity, capabilities (ability to pay) and responsibility, which is said to be common but differentiated.

Chapter 8 reconsiders all of these issues in a new light, that of justification theory, as understood by the French school. This theory focuses on ordinary procedures to resolve disputes and reach agreements. Within this theory, agents of a modern society have to play with a plurality of justification orders, the relevance of which depends on situations of social interaction. This theory helps us to understand which issues should be considered and which norms should be supported as appropriate to the situation in which coordination of climate policies should be implemented.

Chapter 9 concludes by suggesting that climate change is not so much an issue of justice, implying rights, duties and obligations, as an issue of motivation and desire for humanity that may push present generations to make promises that are thrown to future generations, like the people who throw bottles into the sea to send messages to unknown recipients.

NOTE

1. In this book, the expression 'historical responsibility' exclusively refers to GHG emissions produced between the beginning of the industrial revolution (1850) and 1990. The latter date was chosen as a reference for the commitments laid down in the UNFCCC and the Kyoto Protocol and corresponds to the beginnings of organized international action to protect the global climate under a new regime of rules.

2. Whither justice?

In his presentation at a side event dedicated to the issue of historical responsibility during a preparatory meeting for the upcoming Conference of the Parties (COP) to the United Nations Framework Convention on Climate Change (UNFCCC) in Copenhagen in December 2009, Henry Shue of Merton College, University of Oxford, provocatively stated that:

> No one needs academic theorists to tell them what is right and wrong . . . Theorists can make two modest contributions. First, they can answer misguided objections to common-sense principles of right and wrong raised by other theorists and thereby restore the situation to common-sense zero. And, second, they can sometimes systematize some of the plurality of common-sense principles into a coherent whole. (Shue, 2009)

Were it not for the fact that these provocative words were part of an introduction that was intended to be witty, this could be a rather misleading sentence. Taken seriously, it would suggest that basic intuitions without rational thinking can be a necessary and sufficient basis for making pragmatic judgements about right and wrong, fair and unfair, for a complex matter such as international climate policy.[1] I personally consider that raw intuition does not provide a sufficient basis and should give way to a reflective approach paying due regard to logical examination of statements and arguments. The sense of injustice is the most commonly shared, but it is dangerous to rely on the first intuition or resentment. There is more than one step between personal feeling and questioning the general organization of international society. We must appeal to the critical examination of what intuition tells us and make use of reason to reach an appropriate qualification of that what seems intuitively unfair. In doing so, we must consent to the detours taken by arguments. We are then confronted with an intellectual universe full of assumptions, principles, norms, criteria and arguments arranged in different theoretical combinations. Without theoretical work and detailed consideration of carefully weighed arguments, it is difficult to support specific conclusions or statements in a convincing manner.

In his *Theory of Justice*, Rawls (1971 [1999]) defended the method of 'reflective equilibrium' between considered judgements and principles to reach well-established judgements that achieve a satisfying level of coherence. At the beginning of the process, there are beliefs and judgements that result from intuitions or experience. Next, principles, to which I add factual elements, are used to test the justification of intuition-based beliefs and judgements. Consequences of judgements are explored to see if one still adheres to them, all things considered. It is through iteration between the two poles, by placing beliefs and intuition-based judgements in the light of principles and their theoretical and real-life implications, that we can hope to reach well-established judgements. It is by testing intuition that one gains access to considered judgements that one can share with others whose intuitions may differ. Intuition on one side, and reason on the other, are the two poles involved in building relevant judgements on climate justice.

This is a delicate exercise and debates have arisen on determining whether coherence is the ultimate factor of justification (Daniels, 2016), and which pole should command in the end: intuition-based judgements or reason-tested ones, as underlined by Edward Page (2006, pp. 16–18). It seems that many scholars who have developed thinking on global climate justice are keen to give ultimate precedence to their primary moral intuitions, at the risk of torturing concepts and theories to obtain confirmation of these intuitions. In this book, my personal inclination will push me to give precedence to the weight of arguments and to respect the integrity of concepts, even if this leads me to dismiss the intuitions of some.

2.1 WHAT JUSTICE IS ABOUT

2.1.1 The Scope

The word 'justice' reflects a concern expressed by many statements such as 'to each his own' or 'equal conditions, equal treatment'. Before examining this field, its boundaries must be marked. Approaching a problem through the category of justice implies setting aside some alternatives: violence and the strict balance of power on one side; selflessness and altruism in their various forms (sympathy, compassion, friendship, love, and so forth) on the other. With the idea of justice, the goal is to shape a relationship with other individuals that enables social cooperation without bitterness or envy, while avoiding the assumption that people who wish to cooperate with others have a special affinity for them.

The field of justice thinking is shared by various scholars belonging to different schools and disciplines and motivated by different projects. On one side, we find moralists, who are attached to identifying moral duties and obligations that a person should acknowledge and put into practice regarding himself and other people, be they familiar or unknown. They hold a point of view focused on the moral rights of individuals, with a tendency to short-cut features of social organization as contingent and arbitrary elements that can be disregarded. On the other side, we find political philosophers and economists who are interested in the design of social institutions and their distributional effects. The former see justice as a moral concept, whereas the latter consider a politically organized social justice, with concern for rights, but also welfare, freedom or democracy. Naturally, there are considerable overlaps between the areas of interest for both groups.

Jon Elster (1992, pp. 195–200) underlined important foundations of the moral universe of justice, designated as the two principles of 'ethical individualism' and 'ethical presentism'. These mean that a moral justice argument is interested in the situation of individuals and in how various assets are allocated to them or distributed among them. What matters ultimately is the treatment of individuals, not groups, and the past should only intervene if it still weighs on relevant features of their present situation. A political approach to justice will preferably consider the relative conditions of groups within society, for instance income or wealth groups, or the dichotomy between the elite upper class, the middle class, the working class and the precariat, or between natives and migrants. It will also make a distinction between citizens of a given nation state and foreigners.

2.1.2 Essential Distinctions

Beyond differences in approaches, several classical distinctions are essential. There are three main fields of justice: (1) commutative and procedural justice (fair and free-will transactions, fair access to expertise, public debate and decision-making; impartiality in considering the interests of all parties involved in a decision-making process; and so on); (2) corrective or compensatory justice, including restorative justice,[2] for torts and damages unduly imposed upon victims whose rights are violated by certain members of society; and (3) distributive or social justice pertaining to the distribution of material and social resources (income, rewards, honours, property rights) entailed by social institutions. These fields are interwoven to some extent; for instance, procedural fairness is critical to the implementation of corrective justice. These three fields are described in more detail below.

Commutative justice

Here, justice is about fair conditions for carrying out transactions between individual agents and organizations, regardless of the attributes of the partners (wealth, fame, physical or moral personality); its aim is to eliminate unbalanced or unconscionable transactions, from the point of view of the commitment within the relationship (avoiding abuses of dominant positions or removing obstacles to fair access, for example), and to avoid a disproportionate sharing of gains derived from social interaction and participation therein. It is rooted in the values of respect for each other's dignity and reciprocity in the relationship.

Corrective justice

Corrective justice seeks to repair the harm caused to individuals by rights violations by others; its goal is to restore the victim to the full enjoyment of their rights; it may also be aimed at establishing or restoring the broken social link between wrongdoer and victim, to enable the psychological and social recovery of both parties.

Distributive justice

Distributive justice embraces the distribution rules of the main types of assets within a polity; these depend on a set of public institutions, which Rawls (1971 [1999]) called the basic structure of society. Here, the way in which a good is allocated to a person is not judged, nor is a rule taken in isolation, but the overall impact of all the rules and mechanisms governing the distribution of assets among all members of a 'well-ordered' society is considered. Distributive justice has two faces: the contributive one is relative to the distribution of burdens by which individual agents contribute to the production of the common good (for instance, taxes as income taxes or value-added tax aimed at financing public expenditures and redistribution); the other is relative to the distribution of goods and services and bads (pollution, and so on), notably through the provision of public goods and services, protective regulations, the redistribution of income, the allocation of subsidies, social benefits and pensions.

Distribution issues can be approached in two opposite ways. One is consequentialist (a similar idea: teleological)[3] and the other procedural (a similar idea: deontological). Procedural conceptions of justice consider it impossible to say anything about an observed distribution of primary goods if observers have no information about the initial rights situation and the social processes that have generated these outcomes. It is the chains of real interactions between human agents, assessed by the respect of all the rights affected, that will determine the fairness or unfairness of outcomes. For example, according to Robert Nozick's (1974) libertarian theory, it is the legitimacy of the initial property rights and the free and voluntary nature of derived transactions on goods that determine the justice of the end-state of goods distribution, regardless of the statistical differences in results among individuals. In the same category, but with a different basis, are theories that consider 'equal opportunity' as the first and foremost principle of social justice (Roemer, 1998): society should compensate for involuntary disadvantages, and first and foremost those linked to birth and family conditions, to ensure equal opportunities for everyone. Once this compensation has been made, the distributional outcomes are considered fair and equitable, whatever they may be.

At the opposite end of the spectrum, the consequentialist conceptions of justice assume that it is possible to directly judge the fairness of end-results without knowing anything about the social processes behind these results. These theories use external, objective end-result criteria to determine what equitable distribution should entail and to assess the actual distribution according to this yardstick, for instance through comparisons of Gini indexes applied to income or wealth. The proposed criteria differ from one author to another. Note, for example:

- strict equality in the distribution of net benefits or costs;
- a threshold for the maximum income gap between the rich and the poor (for instance, 1:20 at most);
- satisfying the basic needs of all, to ensure fulfilment of the right to human dignity;
- maximizing general economic welfare (standard utilitarianism).

Several approaches combine elements taken from procedural and consequentialist approaches. This is the case of the standard test of 'from each according to his ability, to each according to his needs', since meeting needs is not enough to satisfy the criterion. The same can be said of the Rawlsian theory: once the fundamental freedoms are equally guaranteed to everyone, inequality may be just and equitable under two conditions: (1) equal opportunity is ensured; (2) inequalities sustainably benefit the least well-off social groups. This strange formulation considers the economic processes that affect the level of production within a society, including the effect of the basic structure of rules on incentives to produce wealth to be distributed. Most proposals for global climate disruption mobilize elements from both procedural and consequentialist approaches.

Global justice and local justice

'Global' justice concerns the design of just institutions within a polity, typically a nation state; reasoning then addresses the general institutions of that society. Here, the term 'global' does not refer to issues pertaining to the Earth system. For instance, Rawls's theory is a 'global' justice theory in this sense, though he is not directly considering international issues. Here, general rules and principles are being considered that should cover a great variety of situations in space and time. This is the same for other theories (utilitarian, communitarian, libertarian, and so on) (Kymlicka, 1990). The level of abstraction of these theories is high.

By contrast, what Elster (1992) called 'local' justice concerns the fair treatment of individuals placed in a situation of social interaction (buying goods at the supermarket, signing a job contract, accessing hospital services, and so on) in a society whose institutions do not regulate in detail how to allocate certain goods. It assumes 'local' institutions that have a degree of autonomy in selecting allocation rules. Systematic interest in 'local justice' or 'equity in the small' traces back to the books of Elster (1992) and Hobart Peyton Young (1994).[4] When an issue is tackled as a 'local' justice problem, the assumption is that it should be solved with appropriate criteria considering the nature of the goods involved (dividing a cake between a family, allocating access to organ transplants, and so on) or the identity of the people concerned, while meeting the constraints raised by the existing institutions and the rules of the global polity. Local rules are not primarily compensatory, but relate to empirically specified goods (Elster, 1992, p. 4). However, local justice is concerned not only with the 'commutative' aspects of social interactions (fair treatment of partners), but also with their distributive aspects, at a 'local' and limited level. Beyond that, the specific features of the situation (circumstances) and the specific aims of the organizations involved are important in defining the issue and finding appropriate solutions.

Justice and equity

The terms 'fair' and 'equitable' are almost always used together, which creates some uncertainty about the difference in meaning. They are connected as expressions of two essential but different components of the overall idea of justice. Equity is not the blind outcome of compliance with a just law, but may require an amendment or a shift from a literal reading of the law in order to reflect circumstances or situations which are not described by the law. Among these circumstances, a central role is often given to differences in income or wealth.

2.1.3 Norms of Justice and Dependency on Social Values and Situations

Moral philosophy makes frequent use of fables featuring interpersonal situations to stimulate intuition. Moral reflections on climate justice often proceed in the same way. This method has some attractive features, but it should be used with skill and precision, since fables may be the vector of hidden prejudice or subliminal messages that force conclusions despite the gap between metaphors and the actual situations to which they are then applied.

Using this method, I wish to give an intuitive impression of the dependence of norms of justice on the structure of situations – including the prevailing social values – in which an arbitration should be made. In practical terms, it also depends on the available information – which should be in accord with relevant social values – about these situations. Consider an archetypal case: how should a cake be divided fairly between children?

In situation 1, there is no information about the children, except that each of them will eat their piece of cake with pleasure. Equitable distribution will depend on a procedure guaranteeing either equal shares (equal outcomes) or, alternatively, equal opportunities to have a 'nice piece' if the units to be distributed are not the same size. If equality of outcomes is a must, the cake should be cut carefully (paternalistic solution); alternatively, the rule can be adopted that a child is chosen to cut the cake, while informing him that he will get the smallest piece (incentive solution). If the goal is equality of opportunity, a common solution is to ask one child to sit under the table and to designate the recipients of randomly selected pieces to be distributed (procedural fairness).

In situation 2, the problem is the same as in situation 1, with the difference that the organizers of the cake-cutting know that some children are four years old, while others are 15. Would it not be fairer to consider the different food requirements of these children and to give larger pieces to the 15-year-olds than to the four-year-olds? This argument underlines why equity and equality should not be confused.

In situation 3, the problem is the same as in situation 2, except for one additional feature: some children prepared and baked the cake, while others are only consumers. Is it not necessary to consider the cooks' commitment to the community by giving them larger pieces? This question involves not only the recognition of merit, but also the incentive structure to be achieved in order for there to be a cake to share.

In situation 4, the problem is the same as in situation 3, but the organizer has additional information: some children come from wealthy families and are accustomed to eating very good cake, while others come from modest families and very rarely get to eat cake. Would it not be justifiable to grant a privilege to the children from modest families in order to compensate partially for the hand they have been dealt in the global society, and to enhance their life experience? Proponents of utilitarianism and supporters of repairing inherited injustices could come together here.

In situation 5, the children are now allowed to either eat their piece of cake or to put it aside in order to sell it on the market. What is a fair distribution in this case? Through the possibility of exchange, what is being distributed is no longer a cake to be eaten, but a valuable resource in terms of exchange, which implies the hope of increasing purchasing power. Thus, reference to the physiological needs of each child becomes meaningless. Should the unequal commercial skills of children, their unequal ability to turn their slice of cake into income, be considered? Should a general standard of wealth redistribution be applied?

The same problem, which is rather simple at first glance, can therefore be approached in several ways, all of which embody a sense of justice. The important elements are: (1) social values and institutions; (2) the relevant features of the situations; (3) the availability of information about the relevant features; and (4) the availability of practical devices or procedures to implement a given criterion of justice that is in phase with reflectively approved social values. Several references, which may be combined, have emerged from my short stories to ensure an equitable distribution: impartiality, needs, merit, unequal wealth, equality of opportunity, the incentive to produce, and 'capabilities' – those substantial liberties of choice that support people's autonomous life projects (Nussbaum, 2011). The adequate qualification of situations and the identification of relevant social values are prerequisites that will determine the aptness of the choice of a specific norm of distributive justice.

2.2 TWO FOCAL THEORIES: RAWLS AND UTILITARIANISM

Bearing in mind the background of climate change, the normative question of what is to be done by whom finds two remarkable and very different answers, with Rawls's thinking on the one hand, and the sort of utilitarianism that economists currently practice on the other. I now provide an overview of both approaches and underline some points of the critical debate to which these theories are exposed.

2.2.1 Rawls's Theory of Justice

Rawls is well known for his major achievement: the comprehensive political theory of justice which he worked on for 50 years. It is categorized as contractual, since it follows the traditions of thinking based on a foundational fictitious social contract. It is also categorized as liberal, because it defines a lexicographical structure of principles giving priority to the principle of equal rights to basic liberties acknowledged for every member of the society. This political theory of justice is aimed at identifying the set of principles on which a just basic structure of society should be built.

Several features of this theory have become cornerstones of contemporary debates on justice: (1) the specific method used to identify the structure of principles: in other words, the fictitious situation of the 'veil of ignorance' covering the assembly of supposed founders of a new society as a metaphor of impartiality; the latter are assumed to know nothing about their own particular attributes (sex, race, abilities, and so on, and their conception of the good) or their place in society (class, social status, and so on); (2) the structure of principles placing access to primary goods – those goods that any rational person prefers to have more rather than less of – in second position behind liberties for all; (3) the specific principle of difference, which states that, as regards access to primary goods, inequalities may not be unjust when they serve the interests of the worst-off, provided that equality of opportunities is ensured; (4) the conceptual positioning of the idea of justice as delimited by the 'circumstances of justice'; and (5) the rejection of utilitarianism, a 'sacrificial doctrine' that is essentially incompatible with the principle of equal rights of every person to basic liberties. I now focus on the issue of the circumstances of justice.

Circumstances of justice

Rawls approached society as a cooperative venture for the mutual advantage of participants. The association of the latter is structured by certain rules of conduct designed to advance their good. When well founded, these rules respond to principles of justice that are designed to be shared by any member of society who is rational and reasonable and needs to cooperate with others. In this community governed by rules of justice, individuals become equal citizens connected to others by a civic link, beyond differences of origins, resources and projects. Hence the key idea of reciprocity between citizens in a well-ordered society that is structured by just institutions.

Rawls's theory is based on a fundamental Humean principle about the conditions of relevance for the idea of justice: 'the circumstances of justice may be described as the normal conditions under which human cooperation is both possible and necessary' (Rawls, 1971 [1999], p. 109). The concept of justice can only be used appropriately when defined circumstances, those of social cooperation, are met. Within this framework, social justice is not the ultimate foundational concept, but a derived one; issues of justice arise because, and to the extent that, human beings need to cooperate.

According to Rawls (1971 [1999], pp. 109–110), the 'circumstances' entail objective and subjective conditions. The main objective conditions are plausibly the following:

- Human beings who coexist within a specific geographical territory, which in normal circumstances assumes a context of repeated interactions over time (the idea of physical links and interaction).
- Human beings who have comparable capacities and skills, with none able to dominate others, contrary to situations governed by the violence of power relations (basic similarity associated with the need to make voluntary agreements or to refer to rules that would be agreed upon voluntarily).
- Human beings whose projects and actions are vulnerable to the actions of others, who may oppose their projects (the idea of mutual dependency underlining a common interest in developing forms of cooperation that enable each person to achieve their ends).
- Resources that are scarce enough to make respective projects compete with one another and to prevent any natural harmony of the respective interests, but still abundant enough in the sense that, through cooperation, solutions can be found that provide satisfactory outcomes for all without condemning anyone to deprivation and starvation.

There are also subjective conditions:

- Each person has developed their own conception of the good, which they feel should be considered and acknowledged by others.
- Distinct individual ends, purposes and plans that result in conflicting individual claims on the resources needed to achieve them.
- All people are both rational and reasonable, but mutually disinterested. This means that they are not essentially altruistic; they do not intend to sacrifice their own interests for the benefit of others. At the same time, they consider the presence of others and seek mutually acceptable rules of coexistence and cooperation; just institutions are required to ensure this fair cooperation.

Rawls also gave a historic meaning to the appeal of the 'circumstances of justice' by acknowledging that a well-ordered society cannot exist without appropriate institutions. To sustain such institutions, a society should have undergone enough economic development to prevent its members from being overcome by a severe or acute scarcity of resources affecting basic living conditions. Hence the principle of just savings to sustain just institutions in society. Presumably, when human groups face acute scarcity, the basic liberties cannot be ensured and the dominant social relationship pattern may not be cooperation, but domination and violence. Rawls (1971 [1999], p. 110) concludes that: 'one can say, in brief, that the circumstances of justice obtain whenever persons put forward conflicting claims to the division of social advantages under conditions of moderate scarcity'.

Seeing justice as an emerging issue in the context of ensuring human cooperation is not a consensual point of view. Those scholars, for instance Shue (1999, 2009) or Steve Vanderheiden (2008),[5] who have a moral focus more than a political one generally reject such origins. For them, the moral precedence of justice is foundational; justice deals primarily with the respect, protection and fulfilment of equal natural rights of human beings, independently of the nature and intensity of the social links those people may have in practice. Such natural rights, which are due to every human being from birth, are supposed to be universal, whatever the wealth of societies, and thus also ahistorical.

More specifically, let me mention the arguments of Hendrik Philip Visser 't Hooft (1999), who developed an external and internal critical view of Rawls's thinking about cooperation by considering the issue of intergenerational justice. If justice were the outcome of human cooperation and entailed reciprocity, there would be no rational meaning to the idea of justice between non-overlapping generations, which would go against the intuition of this author. From a theoretical viewpoint, 't Hooft considered that the sense of justice must precede the circumstances under which it is supposed to emerge.

At the heart of his criticism lies the way in which Rawls addresses the issue of temporal dynamics to conceive his principles of difference and just savings, and the link to unborn future generations. First, the aforementioned principles in favour of the best access of the worst-off to primary goods should be understood through incentives to production, savings, investment and economic growth. In this sense, they entail considerable social cooperation. However, with an undefined time horizon of economic development, groups of the least well-off must see their composition changing over time. At a given time, for instance in 2016, the view of the development process in the future will include people who are not yet born in that year. However, it is not possible to say that living people are cooperating with unborn people. 't Hooft adds that in an intergenerational context, the idea of competition for scarce resources can only have substance when living generations decide to take account of the interests of future unborn generations. For this author, it is the intuitive sense of justice that validates this account: scarcity arises because of the sense of justice due to the unborn. Consequently, the sense of justice must precede the circumstances under which it is supposed to emerge. Thus, the most celebrated principles of justice as per Rawls necessarily escape the circumstances of cooperation. This criticism by 't Hooft converges with that of Thompson (2009, pp. 3, 5):

> We cannot make bargains or agreements with the dead or the unborn; it is difficult to understand how dead or future people can have rights or participate in a just process . . . According to contractual accounts, the duties that citizens owe to each other as members of a polity arise out their consenting participation in relationships that they have formed with each other. This is clearly not an adequate basis for a theory of intergenerational justice. An appeal to mutual consent is not, at least in any obvious way, applicable to relationships between the generations.

In terms of logic, arguments that underly the incompatibility between the 'cooperation' approach and the idea of justice between distant generations cannot be refuted. They pave the way to two types of solutions: (1) establishing new foundations for the idea of justice, which differ from the idea of cooperation, in order to enlarge the circumstances of justice and to enable a legitimate use of this concept to cover relationships between living and unborn persons; (2) retaining the connection between justice and the idea of cooperation and reciprocity between living people, with two sub-cases: (2.1) establishing an indirect link to salvage the idea of cooperation between non-co-temporal persons and generations, and (2.2) dismissing all uses of the concept of justice for the link between living and non-living people and finding other sources for the interest that living humans have in the future living conditions of the unborn. I shall return to this issue in the following chapters, but the cooperation approach is clearly the nearest to the situation of international cooperation aimed at building a new regime intended to protect humankind against the threat of global climate change.

The climate threat and the circumstances of justice

Since circumstances are important in determining whether an issue falls within the notion of justice, at least for the Rawlsian school, it is necessary to ensure that these circumstances are met when addressing a global threat such as global climate change. The following is what I find.

Firstly, the global climate is often qualified as a global public good, in the economic sense.[6] These global conditions are preserved or disrupted in a decentralized manner through the behaviour of billions of people when emitting or capturing greenhouse gas (GHG) emissions. In this sentence, several elements are correct: (1) the public dimension is clear since, whatever it is, once the climate benefits or threatens any individual, all others under the same local conditions are affected by the same events, without any device or institution making it possible to exclude anyone who does not pay for access; no one is able to produce their own climate and to keep it for themself; (2) decentralized actions that emit GHGs or capture them in sinks are those that interfere with the global and local climates. But there are strong reasons for casting doubt on the qualification of the climate as a good.

If we refer to Rawls's notion of primary goods, it is defined by the preference of rational human agents to get more rather than less. This definition has no meaning for the climate. If we turn to economics, a good is generally seen as an entity that is useful to humans – technically, it enters the utility or production function of at least one agent – and is both physically isolable and socially exchangeable; to this effect, it is characterized by physical parameters and can be counted using physical units. The population of goods is thus organized through the definition of equivalence classes such that all units in the same class are considered strictly equivalent from the viewpoint of the welfare of agents. This definition is well suited to private goods and should be amended for public goods, which are not exchangeable: what may be exchangeable are not climate units, but possibly quotas of GHG emissions rights, which is very different. However, nothing in the remaining elements can support the assimilation of the global climate as a good.

Local meteorological events (rain, temperature, wind, and so on) are what impact upon the utility or production functions of human agents, not the global climate. The latter is a synthetic scientific construct developed by scientists looking back at several decades of observations of meteorological parameters over a region. These events are not direct variables of utility or production functions – this status is reserved for the goods directly valorized in utility terms – but only belong to the environmental background that influences the conditions of production, that is, determinants of the productivity of basic resources and production factors. In addition, there is no physical delimitation of climate entities that could be numerable and give rise to accounting.

The conclusion is clear: the global climate cannot be said to be a good. It is a rather synthetic way to describe some features of the natural conditions of human life and the economy. Accordingly, we cannot directly apply to the climate issue norms of justice designed for the distribution of goods or for the protection of existing rights to goods.

Meanwhile, protecting the global climate by controlling GHG emissions requires the cooperation of most people in the world because, through the effect of group dynamics – each person being ensured of the actions of others – it would pave the way to an infinitely more efficient approach than isolated actions which, taken separately, have no discernible effect. Even stringent mitigation action by developed countries alone would not be enough to achieve the goal of avoiding dangerous interference with the planet's climate system. Due to the objective characteristics of the situation, strong and sustained cooperation between developed countries, emerging countries and developing countries is a necessity to address the challenge.

The situation will be different if, instead of controlling emissions, the focus is placed on adapting to climate change or on geo-engineering (Hamilton, 2013). Whereas mitigation is aimed at producing global public conditions, adaptation strategies provide only private goods or local public goods and services that benefit only those who commit to action (Barrett, 2008a) and do not depend on significant international cooperation. Geo-engineering presents an intermediate case: if its national interests are sufficiently threatened by climate change, a powerful state – such as the USA or China – could decide by itself to use technological solutions (sending aerosols into the atmosphere, dispersing iron in the oceans, placing large mirrors in space to limit the solar radiation reaching the Earth, and so on) without being technically obliged to obtain the agreement and active cooperation of other states (Barrett, 2008b). This would be a case, as with monetary regimes, in which a leading power is sufficiently concerned by an issue to undertake alone an endeavour that would also concern other countries. Thus, the ideas of justice cannot be the same for GHG reduction strategies and for those involving adaptation or geo-engineering, not only from a practical point of view (the most obvious), but also regarding normative benchmarks, since conditions for cooperation vary significantly between them.

Secondly, a new climate protection regime is mainly negotiated among states and it is from states that major commitments are expected, since they are generally acknowledged as representing the peoples in their overall economic, social and political components. Furthermore, they have the legitimacy and power to enforce the commitments they make. Shaping a climate protection regime clearly belongs to the realm of international relations. Neither individuals nor different types of cross-border communities – social networks, non-governmental organizations (NGOs), business – are central actors in such a regime. At best, scientists and NGOs may exert pressure and obtain influence on some parties, leading to change of some procedures, and conduct indirect action through public opinion. As for business firms or city networks, the sustainability of their pro-climate initiatives depends on the persistence of state commitments.

Thirdly, climate policies are undeniably costly, whatever the controversies about the level of costs (moderate, acceptable or high) resulting from various strategies. At the same time, the scarcity of resources needed to achieve a low-carbon economy is not such a constraint that it was unthinkable or impossible for human beings to keep their global GHG emissions under control at a level maintained under a major climate threat threshold: experiencing dangerous human-induced climate change was not the fate of humankind when this began to be considered as a governmental issue in the late 1980s. Fifteen years later, it was still possible to efficiently address the challenge with the use of existing or imminent technologies, if the approach was both systematic and strongly stimulated and coordinated at the international level (Pacala and Socolow, 2004). In 2016, the race against time was lost. If humankind has moved into the danger zone in an irreversible way, it is not because it lacked the means to avoid this zone; it is because it did not want to take the steps to organize this.

Fourthly, embedded in very different cultures and socio-economic conditions, different countries clearly and legitimately have different ends, purposes and priorities, specifically regarding the importance to be given to climate protection in relation to economic development. These differences result in conflicting plans and claims, which is reflected in the hesitation on the path towards international coordination since 1992.

Fifthly, states – those 'cold monsters' as they are sometimes presented – are clearly not altruistic, but they learned long ago, through repeated experience of interaction with others, that they must try to find a *modus vivendi* with them and even cooperate with them for specific goals.

The conclusion is straightforward but mixed. Because of several of its features, the global climate threat rightfully belongs to the perimeter of the concept of justice, largely because it clearly calls for cooperation between states. However, although international cooperation is valuable, its necessity is not uniform across climate strategies. Cooperation is imperative for stringent mitigation policies to be efficient in terms of their ultimate declared purpose. Nevertheless, it is not a *sine qua non* of all climate policies. Here, motivations and requirements are connected: adaptation and geo-engineering may provide second-order 'alternatives', if not first-rate substitutes, in case international cooperation cannot be achieved under acceptable conditions. Moreover, since the global climate is not a good,[7] to organize the field of climate protection, values and concepts of justice can only be mobilized in an indirect or peripheral way. Ultimately, with the threat of climate change, the issue is not to determine which general foundations to give to a just global society, as was Rawls's ambition for national societies. Here, it is necessary to tackle a field of action framed by the 'huge but limited needs of cooperation' on a specific topic with numerous ramifications for other collective goals, such as economic development and the satisfaction of basic needs. Thinking on global justice cannot avoid being affected by seeing the place of justice arguments being accordingly restricted and bound.

2.2.2 On Utilitarianism

Utilitarianism provides the standard background for normative welfare economics, even if well-known scholars such as Amartya Sen (1977, 1979) have expressed strong criticism of this reductionism. Utilitarianism is a type of consequentialism. To keep things simple, consequentialism means judging the value of an action on the sole basis of an appraisal of the consequences of that action. The opposite is a deontological approach, which assesses the value of an action based on its intrinsic features. For utilitarianism, the goal of society, depending on the interpretation, is to maximize the sum of individual utilities, or of such a weighted sum, or of this sum per capita, or to find the social rules that will produce those outcomes in the long run: the greatest good for the greatest number of people. With crude utilitarianism, no issue can be raised regarding natural rights or the intrinsic value given to liberty, freedom of choice or integrity. For sophisticated utilitarianism, enforcing a system of personal rights may be considered if it is shown to be part of society's rules to maximize overall welfare. Thus, standard economic utilitarianism combines individual welfarism, consequentialism, impersonality (ignorance of issues concerning rights and personal autonomy) and a vision of the general interest of societies as a sum of individual well-beings to be maximized ($Max \sum a_i U_i$) (Sen and Williams, 1982).

One of the main qualities of this theory is that it has lower demands for data than others, since it disregards many features that other theories consider relevant. It also fits the instrumental perspective that is praised by both business circles in relation to mass consumerism and academic circles in relation to the huge possibilities for formal analytical developments. The usual criticisms are that utilitarianism confuses rationality with the egoistic defence of self-interest by every human agent, is indifferent to the moral motivations of agents, fails to acknowledge the intrinsic value of the freedom of choice of people and societies and basic human rights, and thus sees no principled objections to sacrificing some people for the benefit of all others.[8] Nor does it recognize that the social values and goals of a community cannot be reduced to the sum of individual satisfactions in relation to the personal preferences of each consumer. Utilitarianism is then at odds with the contractarian tradition of political philosophy from Jean-Jacques Rousseau to Rawls.

The standard economic view concerning the distribution of private goods is that equity and efficiency are separable: there are as many Pareto optimums as there are initial distributions of income and property rights. Income and property rights can always be redistributed according to social values without interfering with society's capacity to achieve an efficient allocation of goods. For more than 50 years, the practical criterion recommended for public investment decisions in that context is the Hicks–Kaldor efficiency criterion: a project should be supported when the present value of expected social benefits exceeds that of expected social costs, which would make it possible for agents to whom benefits accrue to compensate those bearing net costs; this is the criterion of the potential compensation of losers by winners.

Economic models pertaining to climate change developed on this basis show that a single worldwide carbon price should be set in order to achieve productive efficiency by lowering GHG abatement costs; this could be done through an international tax or a world market for carbon emission quotas (Stewart and Wiener, 2003). Regarding distributive aspects, the non-cooperative game-theoretic framework used (Barrett, 2003) underlines the need to find a rule for burden-sharing that ensures a net benefit for each party (the Pareto approach): no major GHG emitter should be a net loser in the new climate regime. This recommendation was emphatically reiterated by Eric Posner and David Weisbach (2010, 2013). What is presented as a feasibility constraint more than an ethical principle goes against many of the proposals made in the name of climate justice, such as an equal per capita allocation to countries or a burden allocated in proportion to historical responsibility. This exemplifies a Gordian knot of justice and cooperation.

However, within utilitarianism, proposals do not stop there. A given state of the global climate is mainly produced as a by-product of various economic activities. In other terms, as far as mitigation is concerned, one is faced with a case of the decentralized production of environmental conditions possessing some features of global public goods. In that case, some authors (Chichilnisky and Heal, 2000) have demonstrated that, from a welfare viewpoint, efficiency and equity are not separable: there is no degree of freedom as in the case of an economy of private goods.

In the general case, a welfare optimum would be reached when agents pay towards the production or maintenance of a public good only an amount equal to their marginal willingness to pay for the corresponding level of supply of the public good to which they are given access.[9] Due to unequal income levels worldwide, trade-offs between private and public goods differ considerably, as does the marginal willingness to pay for a given level of the public good. Thus, if to achieve productive efficiency a single carbon price is imposed on all countries and then decentralized in their domestic economies, welfare efficiency will be seriously impaired, unless transfers of assets (allowances, finance and technology) are simultaneously made towards low-income countries and people. It is not just for equity but also for efficiency in terms of maximizing welfare that these transfers are required.

Table 2.1 Possible attractors for international climate negotiations

World carbon price	GHG concentration targets	
	Loose (650 ppm CO_{2eq})	Stringent (450 ppm CO_{2eq})
Single	Low transfers	Massive transfers
Differentiated	No transfer	Significant transfers

If such transfers are not envisaged, or if their level is too low, a second-best solution is to set differentiated carbon dioxide (CO_2) prices across the regions of the world. In one of their first papers on the subject, Graciela Chichilnisky and Geoffrey Heal (1994, p. 444) reached the conclusion that:

> It is not generally true that Pareto optimality requires that marginal abatement costs be equated across countries: this is true only if marginal utilities of income are equated across countries, either by assumption or by lump-sum transfers across countries. If richer countries have a lower marginal valuation of the private good, then at a Pareto efficient allocation, they should have a larger marginal cost of abatement.

According to this often-forgotten analysis, the choice of the climate target, the international price regime and the level of transfers are interdependent: they cannot be chosen separately. Several combinations could become attractors for international negotiations, as shown in Table 2.1, which considers four possible levels of transfers.

Even from a utilitarian approach, it is not so easy to determine a first-best, just, fair and efficient solution to the threat of global climate change. This is even more difficult for a second-best approach.

Although fundamentally opposed, Rawlsian theory and utilitarianism have in common the desire to take economic processes into account in order to arrive at a judgement on the justice of a state of society. This equally applies to the relations between co-present citizens of a nation state and to relations between distant generations. This is a key lesson to remember when addressing the issues of climate justice.

NOTES

1. As a matter of fact, there is some doubt as to the humoristic intent of the statement. In his 1999 paper, Shue was already appealing to universal common sense to defend his views on the obligation of industrialized countries to pay for GHG emissions reductions and other sources of global environmental disruption.

2. Restorative justice is aimed at creating a community around victims to help them overcome the harm they have suffered. It also involves direct contact between offenders and victims to help offenders avoid any subsequent offence. It counts on active citizen mobilization rather than just the involvement of institutions implementing abstract norms of punishment.
3. Martha Nussbaum (2013, pp. 470–71) stresses that a teleological approach may include the capability of choice as a component of the good to pursue, with the effect that this approach is more than an instrumentalism serving an externally defined good.
4. Young (1994, p. xii), claimed that 'it is possible to analyse equity in the small without resolving what social justice means in the large'.
5. Steve Vanderheiden (2008, p. 49) writes: 'justice is related to equality, and this typically takes the form of a premise that all persons are morally equal . . . Assuming that persons are moral equals further entails that *ceteris paribus*, no person should be better or worse off than any other . . . arbitrary or undeserved inequality in outcomes is taken to be unjust'.
6. For economists, the distinction between private and public goods is intrinsic to the nature of these goods. It is based on their qualities regarding two criteria: rivalry or non-rivalry in consumption (or usage); and excludability or non-excludability. A typical public good is one that can be consumed simultaneously by many agents without loss for any of them, in such a way that no technical device can exclude some agents from consumption once it has been offered to others. Private goods are rivalrous and excludable. Club goods can be consumed by several agents, but often with limits on capacity (highways, and so on); the key feature is that they can be made excludable (pay television, for instance). These distinctions do not presume the institutions by which these goods will be produced or managed. They say nothing, as such, about the property rights of different agents, but show which types of problems will be encountered by different categories of goods.
7. Note that to capture the stakes of an evolving global climate, the United Nations has forged the concept of the 'common concern of humanity' (Shelton, 2009), without putting forward that of the 'common heritage of humankind', adapted to 'resources' or 'goods'.
8. In the context of a discussion on fair practice in organ transplantation, Elster (1992) echoes a proposal made (seriously?) by a utilitarian to set up a lottery to kill a certain number of people every year in order to take their organs and transplant them for the benefit of ill people. Since, with one body, several organs can be extracted and several lives saved, the solution would increase social welfare by adding a net benefit in terms of life expectancy.
9. For public goods, total demand results from the matching of production costs with the aggregated marginal willingness to pay of all agents.

3. Justice and international coordination

Although there are international institutions on a global or regional scale, these are not equivalent to a state or world government. Existing global governance is comprised of information devices, expertise, consultation, guidance, approval of agreements, partial regulation, standardization and financing, without having the political, administrative, judicial and military means to make a state. For international problems and even global ones, action can only be based on the voluntary cooperation of states, which is itself influenced by the actors and stakeholders on the national scene, communities, businesses and civil society organizations. This has two consequences: (1) the unanimity of states is required to ensure the conclusion of a new universal agreement, even if other rules may be adopted later for the day-to-day management of an organization or an international institution; (2) the rules are only applicable to those who join voluntarily. That said, the sum of previously accepted rules gradually forms a set of benchmarks that frame the behaviour of every state and limit its autonomy, while providing a fall-back position: agreements generally establish the conditions under which a party may withdraw but, in any case, a state may pull out of a convention or a treaty.

In this chapter, I first look at justice from the viewpoint of international relationships by considering how it would be incorporated differently into three basic models of international coordination. In section 3.2, I provide an overview of the various conceptions of the content of justice in an international or global context. This ranges from national-centred conceptions to cosmopolitanist ones. This corresponds to an examination of international coordination from the viewpoint of theories of justice.

3.1 ARTICULATING JUSTICE WITH MODELS OF INTERNATIONAL COORDINATION

The prospect of climate change raises a specific question: how can countries that differ in most of the relevant dimensions succeed in reaching an agreement on the establishment of a new international regime? This question cannot be avoided, since there is no higher global political authority with all the attributes of a government to impose a rule common to all. It is even more pertinent given that the goal at stake is to preserve global conditions that some parties consider to be essential, while others do not; and still more given that the parties express their desire to achieve a fair and just agreement. In fact, the importance of justice in international relations depends very much on the way we consider coordination and cooperation in that context.

In the literature, we find three interesting models that each give a different role to the idea of justice. First, there is the strategy of creating an 'epistemic community' within which both goals and the perception of the action to be launched are shared. A second model is one in which the states are supposed to seek a mutually advantageous regime through an exchange of concessions and commitments, with every nation state defending its own national interests before others. The third model consists in proposing foundational rules for the international community based on what would be the legitimate rules of an ideal counterfactual state that a global, cosmopolitan community would adopt. The latter envisages humankind as a unified political community, over and above states. Each of these models has its virtues, but also comes up against significant obstacles. Each one defines a precise role for the idea of justice.

3.1.1 The Epistemic Community Model

This type of coordination does not formally depend on a process of political negotiation between states under the auspices of international organizations. It may precede it and develop simultaneously. Following in the footsteps of Peter Haas (1990), an 'epistemic community' represents the creation of a network of scientists, experts and decision-makers that transcends national borders, who all share the same views on various aspects: the phenomena at stake and the causal relations at work (even if these ideas eventually prove to be false); the patterns to adopt concerning the action to launch; and the values determining how public and private actors should behave. Where they are positioned within social networks and institutions, members of this epistemic community act simultaneously to influence high-level decision-makers in government, business and civil society. Through this parallel action, they actively help the states to converge on the priorities and design of the policies that each state will implement. Shared knowledge and beliefs lead all actors to adopt the same perspective on the problems and the course of action to be instigated.

To the extent that this type of coordination is not based on formal rules, the questions of the fair sharing of efforts or the distribution of rights are not considered. Comparing situations, flushing out inequalities or making specific claims to obtain compensation in the name of justice are not relevant behaviours. Placed in a rationale of virtue, each contributor does what they feel is required as their duty without committing to the game of comparisons with others. In so doing, they are prompted by the feeling of being part of a larger community of action.

The challenge of this model lies in ensuring successful articulation between the cognitive terrain and that of the values that inform policies and action. Its central rationale is in enhancing the convergence of world views, integrating objective and normative inputs. The coordination of behaviours is expected from this community of views. The model for action that results from it does not acknowledge any sharp discontinuity between scientific knowledge and values, but instead plays on the continuity between them. Actions to be undertaken are unambiguously perceived as self-evident and a 'must' that is objectively determined. Among the different agents, the central character here is the expert, a mediator between decision-makers who engage society, and scientists who speak of facts and natural laws. The principal means of enhancing this model is therefore the creation of committees of international experts who can establish a common set of undisputed data, deliver certifiable scientific statements and give sound justifiable recommendations to decision-makers all over the world.

The IPCC example

Unquestionably, this model has had an influence on the approach to the issue of global warming. The expected role of the Intergovernmental Panel on Climate Change (IPCC) was indeed to stimulate widespread dissemination of the knowledge available and to facilitate convergence on a diagnosis and concerted methodologies to support country initiatives and the organization of international action. The IPCC was established in 1988 by the United Nations Environment Programme (UNEP) and the World Meteorological Organization (WMO). Since then, the IPCC has accompanied the different development phases of a new regulatory regime for the protection of the global climate. Its general mission is to proceed to a critical examination of the published scientific literature worldwide, and to draw from it a pertinent synthesis to enlighten governments and all decision-makers concerned. Led by renowned scientists, the IPCC works directly with hundreds of researchers and experts from different parts of the world, who are chosen based on suggestions made by governments. One of its original features is that its reports are subject to a dual review process, by both peers and governmental representatives from all parts of the globe: the complete report is under the sole responsibility of the scientists, but they are expected to take account of and respond to every comment made worldwide on their drafts; the summary for policy-makers must be approved in detail by the assembly of government representatives. Though it may lead to minor cuts in the reports and some understatements in the wording, this process avoids cheating science through political manipulation and obliges governments to argue on scientific grounds to obtain a slight change in a draft statement they consider inappropriate.

To obtain the trust of other scientists, the public and governments concerning the outcomes of such expertise, it has proven desirable to ensure a satisfactory representation of the main geographical regions of the world through the choice of participating scientists, and to maintain the diversity of scientific disciplines involved. This search of balance should not compromise the quality of the scientific syntheses or the soundness of proposals made. Thus, quality and rigour in the procedures for writing, commenting and correcting draft reports are key determinants. Of special importance is the obligation that IPCC authors have to give a consistent answer to every comment they receive from reviewers. These elements converge with the requirements of reciprocal respect at the heart of commutative justice.

Coordination variables

In this model, the coordination variables are comprised of scientific data and statements; methods used to organize and synthesize information; valid norms and procedures, for example in modelling or collecting data; and finally, the means and conditions for making sound recommendations. The intrinsic efficacy of this approach is obviously difficult to measure. Nonetheless, rigorous procedures and widespread critical exchanges and discussions, with clearly defined roles, enable it to come close to the ideal of objectivity and fairness expected of scientific practice. Here we can see a necessary condition for international coordination on issues with a strong scientific background, such as climate change.

Drawbacks

Communities of views and mutual trust are clearly important prerequisites for ensuring that this coordination mode functions. These two conditions are generally not met by the whole society in pluralistic democratic societies, and even less so on the international scene. Instead, epistemic communities remain partial coordination vectors among transnational groups. They are real facilitating factors in ultimately reaching a formal international agreement, but are not a sufficient condition for full international cooperation.

IPCC contributions have proved insufficient to trigger timely international action to protect humankind against the climate threat. Uncertainties, scientific on the one hand and intrinsic and historical on the other, have been raised to justify a wait-and-see attitude. They have been used either by dissident scientists or by actors whose interests would be threatened by widescale action to control GHG emissions. In this regard, scientific expertise on climate change has not enjoyed especially favourable treatment compared with that of any other contemporary situation of risk to health and the environment. Commutative justice at the heart of the way in which science operates collectively has been undermined by unfair attacks due to vested interests, and confusion that many businesses and political actors were keen to utilize in order to downgrade the urgency of climate action. Opponents of climate policies have continually attempted to cast doubt on the merit of aggressive protection against the climate threat, which as time goes by would require more and more stringent efforts to cut greenhouse gas (GHG) emissions at the same time as financing investment in adaptation. These opponents have also undertaken to form alternative epistemic communities, which could compete with those that have been created since the establishment of the IPCC, and so feed the social representation that the science is controversial and inconclusive.[1]

More generally, scientists have lost their past social authority. Numerous personalities (company managers, public officials, important names in the scientific community, and so on) consider it legitimate to substitute their own judgement of the reality of climate change for the picture presented collectively by competent scientists. Many are ready to give more credit to any 'naysayer' who receives media coverage, although these individuals have generally removed themselves from the reign of the scientific community's rules of critical discussion and publication.

Thus, however rigorous the procedural rules of scientific expertise, the IPCC experience exemplifies the difficulties involved in forming epistemic communities. Take the so-called 'Climategate' controversy that began a few months before the 2009 Copenhagen Conference.[2] This scandal targeted the suspected unscientific practices of Phil Jones, a prominent climate researcher, and head of the Climate Research Unit (CRU) at the University of East Anglia, a man who devoted most of his time to setting the time series of the temperature record across the world. Then came the discovery in 2010 of two factual mistakes in the 2007 IPCC Working Group II report, which led to a huge IPCC-bashing campaign, extravagantly rebutting all IPCC work as an ideological endeavour conducted by activists (Pearce, 2010).[3] Genuine and rigorously led scientific expertise is not enough to avoid interest-based strategies by social actors to destabilize and delegitimize climate science, with the effect of undermining its benefits in terms of collective action.

Finally, a clear disconnect has emerged between the state of scientific knowledge on climate change and the choice of a given climate policy: the scientific picture obtained has not itself identified a precise attractor or a well-established threshold that would naturally impose itself with the force of evidence. Despite the rather old focus on the 2°C target, no precise goal for action within the next 20 years or so can be automatically derived from the current state of knowledge. Consequently, the assumed continuity between knowledge and action has been systematically refused, as shown by the reality of the policies implemented and envisaged. These policies show a huge gap with the scientific diagnosis of what should be done to reach the goal defined by the United Nations Framework Convention on Climate Change (UNFCCC), even after the 2015 Paris Agreement.

To sum up, the key feature of justice involved in the epistemic community model is neither reparative and compensatory, nor distributive. It is a justice of virtue concerning the interactions of the main social actors involved in the climate game. Experience linked to climate change demonstrates that what is required is the formation of a solid ethics of expertise, combining an ethics of responsibility regarding the consequences of public choices and an ethics of discussion regarding the conditions under which expertise is organized between its 'producers' and 'consumers' and used by them.

This raises an ethical challenge for different stakeholders: the scientists involved in expertise activities, but also scientists who conduct research with a view to direct technological applications; organizations (public institutions, agencies, authorities, non-governmental organizations (NGOs), businesses) that conduct or commission expertise activities; local and national government officials or sponsors who are the recipients of expertise; companies whose activities, products or techniques have to be reviewed by experts with a view to assessment and regulatory decisions; associations and other civil society groups that are concerned by subjects covered by expertise and that ask to take part in this expertise; and judicial circles which, referring to the results of expertise, derive normative conclusions about the legal rights and obligations of individual actors.

It is the whole social ecosystem of expertise that would have to reconsider positions, roles and relations from the viewpoint of fairness. This includes the respect due to each role and self-restraint in the desire to go further than the capabilities attached to roles can legitimately support: scientists are not elected and do not legitimately represent the polity, while governments and businesses are not scientists and have no legitimacy to give their own opinions on the natural processes involved in climate change. But respecting each others' roles should not prevent any cross-questioning between people with different roles; in fact, such cross-questioning should be encouraged and systematized in order to facilitate mutual understanding and to upgrade the quality of expertise as a social function acknowledged to be critical for the quality of public choices.

3.1.2 The Model of a Mutually Advantageous Agreement between Self-Interested States

The preferred model put forward by economists is the one in which states endowed with well-constituted interests – a utility function – freely negotiate explicit new rules to govern their relationships or the use of certain common resources, or to promote common activities. They proceed to exchanges of mutually advantageous commitments regarding their respective interests (Barrett, 1991, 2003). Concerning public goods and natural conditions to which actors previously had free access, abandoning these to the unilateral, self-interested initiatives of each party could only lead to their destruction, or to a very insufficient level of production or conservation – 'the tragedy of the commons' (Hardin, 1968) – due to structural incentives to free-ride. This is a typical case of conflict between private and social benefits. International cooperation is aimed at overcoming this destructive rationale.

Considered as an isolated issue, coming to an agreement on a climate change regime should enable each country to obtain a net advantage compared to a non-cooperative equilibrium and, specifically, to the previous international relations situation.[4] No country should be made worse off or feel itself to be a net loser in a proposed agreement. This is what Posner and Weisbach (2010, 2013) put forward as 'International Paretianism'. In the opposite case, a country perceiving itself as the potential loser would prefer to remain outside of the agreement, as with the USA for the Kyoto Protocol and possibly the Paris Agreement. In that, Posner and Weisbach say they do not see an ethical principle, but a political feasibility constraint, whereas they appropriately acknowledge that it could also be defended as an ethical principle. The latter case can be defended when one gives an intrinsic moral value to the existence of diverse sovereign democratic states allowing citizens to elaborate their own rules of coexistence, with the outcome of enhancing valuable diversity in human political organization and culture.

In line with Scott Barrett or Posner and Weisbach's analytical framework, the feasibility constraint is still more stringent than these authors say. The potential signatories to an agreement should find no greater interest in staying outside the agreement if the latter can stand and be implemented by others. Indeed, when a public good is at stake, a country may have a double interest to remain outside of an agreement, firstly in benefiting from the cooperation of others without assuming a share of the cost of the action, and secondly in taking additional economic advantage of collective action by others. For that, it may commit political resources in helping a coalition to form in which it will not take part. For instance, it has been said that the interests of the USA would have been best served in the 1990s and 2000s by a strategy in which this country: (1) remains outside of a climate coalition, and so avoids the cost of emissions abatement and of fast changes to its energy system; but at the same time, (2) actively contributes to persuading various countries to enter the climate coalition, for instance by supporting climate science (modelling, satellite observations, and so on) and making use of various sources of influence on other states; and (3) pushes domestic innovations in non-carbon energy sources in order to sell them to the countries belonging to the climate coalition that are less capable in the field of technology.

Obviously, if each party tries to benefit from the cooperation of others without contributing itself, the whole construction unravels before it even has a chance to take shape, and this leads back to a non-cooperative equilibrium. Therefore, taking this second constraint into account is as imperative as satisfying the first: it is important that countries joining a coalition for the protection of the global climate gain more than countries staying out of it; they may need a little help to this effect (side-payments of different sorts for insiders, and taxation and penalties for outsiders).

Within this model, the parties most anxious to see international action succeed are requested by others to offer different forms of compensation so that the latter decide to join a coalition capable of tackling the problem. This was Russia's game between 1997 and 2004: asking for various types of compensation before agreeing to ratify the Kyoto Protocol. Through the offer of compensation or subsidies to potential new entrants, this model assumes a process of progressive enlargement up to an equilibrium at which incentives to be in and to be out of the coalition are equal. Since this process reaches a limit beyond which insiders would be better off as outsiders, we cannot generally expect the coalition for a public good to be universal, unless there is a strong coordination stake within the game,[5] or a strong threat of credible retaliatory measures against outsiders.

Coordination variables

In this model, the respective behaviours of each country constitute the variables on which cooperation can stand. What each state is prepared to 'buy' is the contribution brought by a specific behaviour by the others; what each state is willing to 'sell', under profitable conditions, is its own commitment to a specific behaviour. In the case of climate change, two variables deserve consideration: respective national commitments regarding GHG emissions levels, and commitments to transfer financial resources or specific rights to other parties. In the past, to obtain sufficient support for the Kyoto Protocol, a generous allocation of emissions quotas to certain countries (Russia or Australia in the global context, Spain in Europe) has played a key role in convincing certain countries to join the coalition.

The allocation of emissions quotas resulting from such negotiations cannot fit a criterion of distributive justice understood *in abstracto*, be they past emissions, equal per capita allocation or the same percentage of cost impact on gross domestic product. When the negotiation undertaken is more comprehensive, without being limited to one issue such as climate change or, within climate change, to the allocation of the carbon budget between states, more complex compensation schemes may arise by organizing a game of concessions across several negotiation scenes (Keohane and Victor, 2016). It is then possible to initiate a sort of bartering process in which a country accepts obligations relative to the climate problem, for example, in exchange for advantages offered in other areas that relate more closely to that country's concerns.

The way in which this agreement model functions depends on one important condition. As with any exchange model, the country that acquires a 'good' must have a strong guarantee as to the quality of the good acquired. In practice, each party must be assured that the agreement is and will continue to be effectively implemented by the other actors. What can be done to ensure they fulfil their promises? For climate change, states must have access to reliable measurements of the GHG emissions of all other states; they should also have the legal and practical power to implement incentives and sanctions against states that do not comply with their commitments. The whole negotiation process based on state interests intrinsically implies that states compare respective actions and benefits obtained. This intrinsic logic of comparison provides the objective basis for the emergence of complaints about injustices.

Here, we are faced with a type of issue that is intermediate between commutative and distributive justice. It is commutative since it concerns the horizontal relations between agents linked by a common set of rules. At first sight, sustaining the mutual respect of partners and self-esteem imply that each party is determined to fulfil the commitments they made; this is typical of commutative justice. But the very moment of comparing the costs and benefits to be borne by each party will inevitably open the Pandora's box of claims about distributive justice. The different parties may ask whether it is equitable that they should bear such a level of costs while others appear to have far less to shoulder. Debates on criteria of distributive justice cannot be avoided in this context. Any situation implying the need to process comparisons will result in such outcomes and pave the way for discussions and objections about distributive justice.

Another key feature of this model is the presupposition relative to the legitimacy of the initial pre-agreement context, since it serves as a basic counterfactual for all parties: if they do not cooperate through the proposed agreement, then they return to the previous situation. When establishing new international relations, adopting this counterfactual as a benchmark to assess a proposed agreement may be seen as practically self-evident, as with Posner and Weisbach. But when the issue of distributive justice is on the table, this choice can be and is regularly called into question by some parties, based on their own conceptions of justice.

Here we find the core tension that affects this model between two concepts of justice. On one side, we find the tradition initiated by Hume and enhanced by Rawls which, through reflection on the 'circumstances of justice' (see Chapter 2), derives the relevant field of justice from the more fundamental concept of cooperation between human individuals. This may be a one-time interaction or a context of multiple, diversified and repeated interactions, as is the case within a family or a local community. Whatever the type of relations involved, the counterpart of this cooperation-dependent concept of justice is that no issue of justice can be raised in situations where agents do not voluntarily interact and cooperate. Failure to come to an agreement on a specific topic such as the construction of a new regime of action for the protection of the global climate will invalidate any further claim about climate injustice and obligations to restore climate justice.

On the other side, we find an extensive concept of justice, seen as normatively foundational for all social interaction situations. As reiterated by Visser 't Hooft (1999), for instance, introducing an argument of justice implies an act of critical distancing from the real-life conditions of social interaction marked by all imperfections and contingencies of inequality, injustice and exploitation; justice reasoning is based on ideal counterfactuals defining rights and duties. Thus, an argument based on this concept of justice cannot accept as a given the constraints of feasibility and participation that Posner and Weisbach and game theorists underline to support their Paretian model of international cooperation. To enter a relationship aimed at testing the possibility of cooperating on a given problem of common interest automatically generates the conditions under which some parties will make claims against existing injustice in the initial situation. These parties will ask for this injustice to be considered and settled within the new rules of the game at stake, to ensure a sound cooperation between all parties, even if claimed injustices are not directly linked to the topic under consideration.

Under such conditions, it is no surprise that cooperation may fail. In this regard, putting forward justice claims does not improve the chances of cooperation, but adds a new potential cause of failure: if initial positions are claimed to be unjust, a positive outcome for negotiations cannot be accepted if it does not include reparation and compensation measures for the initial injustice. However, the said inherited injustice may be such that the overall benefits of cooperation on a specific topic would not be enough to provide appropriate compensation for the claimant parties. Consequently, states will go back to the situation that prevailed prior to the attempt at an agreement, despite the claimed injustice that is attached to it. The issue raised by the threatened global natural conditions will not be addressed. This is a losing game for all, except for those whose main objective was to gain an audience to ensure that the injustice they perceive is heard, if not acknowledged, by the international society.

Obstacles

For global environmental issues, this model comes up against some specific and well-known difficulties. The first is the absence of coercive measures obliging a country to honour its obligations. In the context of implementing the Kyoto Protocol and of international emissions trading, the urgent proposals made by experts to establish a system of financial penalties for defaulting states were not adopted. Instead, those states were only threatened with losing the right to take part in the international carbon market and to be assigned additional abatement targets in the next commitment period. However, since caps for the next period had not been set, the incentive to comply did not look very strong. As Thomas Schelling once put it, who is going to make war against the USA or China because of their excessive levels of GHG emissions? At the international level, under the present United Nations (UN) rules of the game – there is no 'Environmental Security Council' capable of taking strong measures against ecologically rogue states – climate action can only be voluntary, a reality which has been institutionally accepted since the 2009 Conference of the Parties in Copenhagen.

This considerably undermines the idea of organizing international action based on mutual commitments, and still more that of counting on an international carbon market: how is it possible to ascertain the value of that which is traded if states do not feel bound by quotas they have received or quantitative commitments they have made? States cannot even rely on threats of retaliation by withdrawal from the agreement. Of course, a country that is dissatisfied with other parties' behaviour could choose to renounce its own commitments. This would exacerbate the damage suffered in the long term by all countries. But this solution has two drawbacks that harm its credibility: it is self-punishing at the same time as it punishes defaulting countries, and the punishment will come much later, as will the damage itself.

Furthermore, the controversial and uncertain nature of climate change makes highly speculative any assessment of the future local damage that a country would face. Given the unknown and long-term dimension of most of the possible damage, governments are unable to determine their own interests decisively and thoroughly. On the other hand, mitigation costs would be experienced in the short and medium term, encouraging the mobilization of various pressure groups against any preventive action. Undoubtedly, this asymmetry impedes a comprehensive perception of the well-understood national interests of each country.

Since environmental benefits appear uncertain, the question of the economic costs of mitigation becomes paramount. Focusing on these costs in the context of negotiations leads to their strategic overassessment; it is in each country's interest to represent the costs it would have to bear at the highest possible level in order to justify its demand for a smaller share of the collective effort. But this strategic bidding over cost assessment makes it difficult to subsequently justify a commitment to mitigation, which would appear as a very expensive investment for an uncertain return.

On top of these difficulties, this model of thinking on international cooperation does not explain the original formation of the core group of countries that pool efforts and commitments for the public good. Analysts must introduce exogenous hypotheses, contrary to the basic axioms of behaviour on which the model is based: in other words, that the parties are guided only by strict self-interest. But the smaller the coalition, the costlier the commitment to it will be. The creation of a small coalition has the effect of imposing relatively high economic costs – including a loss of industrial competitiveness – on its promoters for an action that has no real direct effect on its own goal, whether in terms of economic well-being or employment. The initial commitment can only be rationally justified by the prospect of the collective ability of these first-movers to attract other countries to join the agreement in order to reach an environmental efficiency threshold, which is a speculative venture. Therefore, to return to Barrett's (1991) formula, environmental agreements are found most often where they are not truly needed – countries take on commitments that are close to what they would have done unilaterally in any case – and there are none where they are badly needed.

Finally, and this is the most crucial observation for the subjects discussed in this book, nowhere in the model is there room for establishing a distributive solution inspired by the extensive concept of justice. An international climate protection system would have no meaning if the countries that contribute the most to creating the problems do not take part. This is the great 'power of the polluters', which is perfectly illustrated by the US government's decision not to ratify the Kyoto Protocol, and China's repeated refusal, before the 2015 Paris Agreement, to make significant quantitative commitments to GHG emissions abatement. In the interest-based model under consideration, it is up to those who most want preventive action to convince others to join in international action. Some economists have drawn logical lessons from such a situation, formulating what they call the 'victim pays' principle, a reversal of the classical 'polluter pays' principle. It would be up to the net winners from prevention (those who are expected to be the main future victims of pollution) to propose attractive compensation for the net losers from prevention (those who are, above all, the biggest polluters) in order to curb their pollution flows.

This cynical formula, drawn from an apparent realism, is the provocative expression of a strong rationale at work. But it wreaks havoc on any intuitive sense of fairness and is not, therefore, as subjectively realistic as its appearance might suggest. Although victims cannot force polluters to participate in collective action and reduce their pollution, this does not imply that they are always willing to pay these polluters to obtain such a reduction from them. They may prefer to take a stand, denouncing the injustice of the situation and punishing polluters at any given opportunity. As Yoram Margalioth (2013, p. 504) writes:

> We know from experiments, such as the Ultimatum Game,[6] that people are willing to incur significant costs to punish people who they think treated them unfairly. Developing countries might be willing to suffer the harsh consequences of climate change that could have been avoided if a global climate change mitigation scheme had been employed, merely to punish the developed countries for treating them unfairly, even if the global mitigation scheme would have left them better off in an objective (cost–benefit analysis) sense.

Because of the described tension between the two profiles given to the concept of justice, the 'mutual interest' model is intrinsically unstable. On the one hand, in promoting the comparison of due efforts and benefits across countries, not only does the justice or injustice of the burden-sharing solutions emerge from the outset as a central problem, but also the justice or injustice of the initial positions of states. On the other hand, its internal rationale does not provide the conceptual resources enabling it to respond positively to these denunciations of background injustices.

3.1.3 Humankind's Imaginary Institution as a Single Political Community of World Citizens

Historically, international law is primarily based on the accumulation of specific bilateral or partial agreements that are added in successive layers. Such agreements have been negotiated based on a rationale of mutual advantage implying symmetry and compensation between parties. In that respect, they look like the vision expressed in the previous model. However, international law tends in a minor mode to acknowledge another source of obligations and rights alongside the interests of nation states. First, we had the emergence of human rights on the international scene. These define the universal rights acknowledged to every human being irrespective of their nationality, simply for being a human. In that sense, the Universal Declaration of Human Rights adopted in 1948 by the General Assembly of the UN was followed by the emergence of economic and social rights and environmental rights.

Later, the reference to 'humankind' as a subject of law emerged in the 1980s, with the idea of provoking a genuine reversal of the role of states.[7] This was expressed in the concept of the Common Heritage of Mankind, initially in the UN Convention on the Law of the Sea and later within other instances, such as the United Nations Educational, Scientific and Cultural Organization (UNESCO) programme dedicated to historical monuments and sites. This concept expresses the idea that some natural or cultural elements are beneficial to humankind and should not be used, consumed, exploited or managed for the sole interest of certain individual agents or states. Accordingly, they should obey common values, principles and rules. This top-down source of legitimacy would impose these regulatory elements on states, not as elements freely chosen by them based on their own interests, but as obligations imposed on representatives who must act in the name of the interests of a legal entity of a superior rank (Kiss and Shelton, 2000). This partial reversal in the foundation of norms redefines states' responsibility to the international community. States should answer for the way in which they carry out their mandates in the name of humankind.

Although these views are still not dominant, they tend to shake the foundations and standard exercise of state sovereignty and for this reason have encountered stiff resistance, which has been observed in the climate negotiations. This has not halted the movement, and another very similar concept has been introduced into legal texts on climate change – the UNFCCC, Paris Agreement – that of the 'common concern of humankind'. This concept is intended to address issues without any specific spatial dimension, contrary to localized resources such as those found in the deep seabed. 'Issues of common concern', Dinah Shelton (2009, p. 83) writes, 'are those that inevitably transcend the boundaries of a single State and require collective action in response'. They call for effective international cooperation and imply determining a strong institutional basis capable of delivering effective action and compliance in addressing a global problem.

Many analyses in the field of applied ethics and political philosophy relating to global issues, and proposals made by the NGOs concerned, have taken steps on these concepts to develop principles that would shape a politically unified global society that would be organized according to their ideal norms of justice. They then transfer to the level of 'humankind' the concern expressed by Rawls (1971 [1999]) for a 'well-ordered society', conceived to foster the good of its members by implementing a public conception of justice. Conceived from this perspective, an international regime of climate protection would be aimed at instituting the preservation of the global climate as a superior interest of humankind, one that overrides the interests of sovereign nation states. It would also affirm the precedence of universal norms of intergenerational and intragenerational equity over more specific national values and standards. All this cannot fail to transform the scope of states' rights and obligations.

Evidence of these conceptual and legal innovations can be found in the introduction of notions such as 'ecological debt', 'historical responsibility', the 'states' fair share of common resources' or the 'equal right of every human being to dispose of the atmosphere', to name the most well known.[8] These concepts will be discussed in the following chapters. They correspond to an extensive concept of justice, of the foundational type, which is well beyond the limits and contingencies of specific cooperation situations. They embrace within one construct many types of justice: commutative, reparative, contributive and distributive, within an extensive vision of the duties of all people derived from obligations to respect, protect and fulfil human rights all over the Earth. Those rights should be considered and respected without any differences linked to time and space, specifically between living and unborn people and between co-citizens and foreigners.

Coordination variables

Here, full cooperation is assumed. Its variables are the principles (for instance, the UNFCCC affirms the principle of the 'common but differentiated responsibility' of states), the rules of procedure (for example, providing fair access to international negotiations and strengthening the capacity of each party to develop its own expertise), and attributive rules (for example, a pro rata allocation of emissions rights among states according to their population).

The potential advantage of this model is that it focuses on the rules organizing relationships between the 'citizens of the world' and, subsequently, between states. The approach could be less sensitive to scientific uncertainties about the ultimate climate damage feared or the ultimate aggregate level of effort to devote to this issue: whatever the level of efforts to be achieved, if the sharing rules are sound, they will remain unaffected by the contingent reassessment of specific variables.

Drawbacks

This morally grounded approach nevertheless encounters some major obstacles: the absence of an agreement on the common good to pursue or the hierarchy of the main types of goods; the coexistence of contradictory ethical views and claims in the name of justice; the inability of the negotiation context to produce fair rules, since it does not benefit from the Rawlsian 'veil of ignorance'; and the huge gulf between the utopian normative system of reference (a unified humankind wholly organized according to unique principles of justice) and the actual political–institutional organization of the world today.

Beyond the specific discussion of each proposal made within the framework of the cosmopolitanist approach, the theoretical issue raised by this model concerns the relevant distance that a theory of justice should have regarding the situation to which it is to be applied. One approach may be to assess the justice of such a situation with the rules embedded in the social order from which the situation arises. But what should be done if these rules are considered unjust, as is now the case, for instance, in the United Kingdom regarding the past criminal conviction for homosexuality that sentenced Oscar Wilde to prison and Alan Turing to chemical castration and suicide? It is quite understandable to seek meta-norms of justice that should be both more deeply 'just' and still relevant for the existing society. However, introducing alternative concepts of justice may go beyond providing a better interpretation of society's deepest social values, and lead to critical discourses of the very foundations of the present social order. In the latter case, it becomes a matter of designing the general structure of a new model of society and not only of identifying a just solution to a specific issue or conflict.

In this respect, we may wonder whether the model of humankind as a single political community is a deepening of the values at the heart of the existing international society, or the description of an alternative model. This alternative may or may not appear desirable, but does not fit as a tool for finding just and equitable solutions to building a new international regime for the protection of the global climate in today's world. It could determine what is just in this alternative world, but not what is just in our world, unless we want to use norms of justice to transform our society into the alternative world. At the heart of this question is the selection of key attributes to describe the society to which the norms of justice are to be applied. Typically, historical time and the organization of humankind into distinct sovereign nation states are considered as irrelevant contingencies for 'humankind as a unified polity', whereas they are permanent features structuring our world and conditioning any argument about climate justice.

This examination of three basic models of thinking on international coordination and cooperation leaves us rather dissatisfied, although each one presents specific links with the issue of justice. The first, the epistemic community, can be at best a partial model, since it must be supplemented by other motivations and processes; it primarily mobilizes the rationale and motivation of commutative justice for the mutual respect of roles and positions around scientific expertise. The second model gives priority to the acknowledgement of feasibility and participation constraints that limit the legitimate scope of considerations of justice. However, its apparently solid foundations – mutual interests – are threatened by the difficulty of defining national interests in the case of climate change, and by the denunciation of the injustice of initial positions, which is triggered by comparisons that each party makes to determine whether they are a loser or a winner of proposed agreements. The third model, 'humankind as a unified polity', proposes concepts and norms conceived to fit the alternative world that their proponents imagine, but is based on assumptions that are at odds with key features of our present world. Hence two questions: (1) Does the proposed ideal for humankind describe an ideal we should want to join? and (2) Are these norms valid for determining what is just in the real world today, even when viewed as in transition towards the ideal?

3.2 BETWEEN INTERNATIONAL AND GLOBAL JUSTICE

The scope of positions in normative thinking about global justice is very broad, ranging from total rejection of the relevance of the idea of distributive justice in an international context, to attempts to build a cosmopolitanist theory of global justice, through intermediate positions giving more attention to actual international practices.

3.2.1 International Relations without Distributive Justice

For some philosophers who focus their thinking on the political conditions of democracy within nation states – for instance, Thomas Nagel (2005) and, to a lesser extent, David Miller (2007) – the idea of distributive justice loses its direct meaning in an international context, as 'the world' does not refer to a well-established political community with state institutions. For them, strong requirements of distributive justice in an international context lack any moral and political significance. Claims of distributive justice can only make sense and be implemented within the public institutions of a sovereign state or other political community through an appropriate legal order benefiting from the means of enforcement. Justice is an emerging property of human relations when they are organized by a democratic state in an appropriate manner. The idea of distributive justice gains meaning only between people linked by a strong political pact, including the possible use of coercion by public institutions to modify a given distribution of goods and given property rights. It is the democratic order and the political pact established within a political community that transform a 'population' into a society of 'equal citizens' who are acknowledged to have rights that fellow citizens should respect and that possessors can defend by claiming justice in case rights are violated. Consequently, the concept of distributive justice cannot be applied directly to either pure private choices or relations between citizens of different societies.

For these philosophers, states may nonetheless have duties regarding solidarity and assistance towards the poorest worldwide, those whose very survival is threatened or those deprived of freedom. These duties of solidarity and support, or the obligation to acknowledge and provide reparation for past wrongs (Miller, 2007), are outside the realm of distributive justice, but belong to a classical concept of the philosophy of law, that of the 'duties of charity', which Stéphane Chauvier (2006, pp. 89–91) suggested renaming the 'duties of humanity' in modern terms, to underline that such duties echo the universal natural rights of every human being. Thus, dismissing the idea of international distributive justice does not imply adopting a cynical, egoistic view of international relations and demonstrating a lack of concern for misery and poverty or threats to survival: other sources of moral concerns, which are not linked to reciprocity, may be mobilized.

3.2.2 International Justice and Rawls's Law of Peoples

From a normative point of view, international relations were traditionally addressed within the framework of what is known as the 'Law of Peoples', which would translate in modern language into the 'Rights of Peoples'. Here, the basic unit of political reasoning is not the individual person, but representatives of political communities that are conscious of themselves and self-determined, mainly incarnating states.

The first postulate is that of equal sovereignty of states, as indicated in the United Nations Charter. According to Chauvier (1999, p. 21): 'This principle says that states must treat each other as free and sovereign people, who have equal rights, that is who have an equal right to freedom and sovereignty, whatever their size or their strength.' To ensure peace and stability, states must respect the sovereign choices of others. The rules of these good neighbourly relations are what constitute international justice.

BOX 3.1 PRINCIPLES OF RAWLS'S LAW OF PEOPLES

A. Usual Principles

1. Peoples (as organized by their state) are free and independent, and their freedom and independence must be respected by other peoples.
2. Peoples are equal and are the authors of the agreements they make with others.
3. Peoples have the right to self-defence, but not the right to initiate war.
4. Nations must observe a duty of non-intervention, except towards societies in endemic disorder.
5. Nations must respect the treaties and commitments made.
6. Peoples must observe certain specific restrictions on the conduct of war (supposedly motivated by self-defence).
7. Peoples must respect human rights.

B. Other Principles

- Peoples should trade fairly.
- Peoples must jointly establish what are fair procedures for their cooperation.
- Peoples owe assistance to each other when facing natural disasters and in meeting basic needs.
- State governments are responsible for regulating the size of their population; they should maintain the integrity of the natural environment under their sovereign control and its capacity to reproduce by itself.

Source: Based on Rawls (1993a, pp. 46–8).

This was the path taken by Rawls when he strove to identify the principles of such international justice, which cannot be the same as those he established for a nation state (Rawls, 1993a) (see Box 3.1). For this purpose, he changed the terms of convening the participants to a fictitious constituent assembly under a veil of ignorance. Here, there are not individual persons joining an assembly to found a political society, but representatives of sovereign states wishing to set the code of international conduct. These representatives do not have to define the principles of distributive justice governing each state – this should remain a matter of self-determination for each sovereign state – but only those principles that would allow each state to develop its own institutions to ensure its own stability and preserve its future capacity for self-determination.

Rawls wanted to elaborate principles avoiding violation of the pluralism of political choices inherent in an international society of sovereign states. These principles should govern relations between states, even if they are not all democratic, but include some hierarchical–authoritarian states, inasmuch as the latter still respect basic individual rights. They include no requirement for the redistribution of natural resources or economic wealth in the name of a distributive justice imperative. The reference to human rights concerns minimum thresholds for livelihoods and freedoms. Compliance with these thresholds would enable individuals to meet their basic needs and gain real access to the constituent rights of human dignity and citizenship, which are the internal political condition for state sovereignty.

The absence of a redistributive dimension in the relations between states has been the most controversial aspect of this theory. Many commentators believed it is too minimalist. Yet, Rawls's refusal logically took place in the context of the 'ethical individualism' proposed by Elster. Since distributive justice is about state institutions distributing primary goods between individuals, interstate relations are of a different nature. Their purpose is to generate and sustain links between nation states, in other words collective political entities. The relational framework established between states always mediates relations between members of different political societies. This is true whatever is at stake, be it emigration, food safety, financial transactions or communications. This refusal is also rooted in the respect for the political self-determined choice of each well-formed state. Some peoples may prefer values such as friendliness, natural wilderness or the availability of free time to economic wealth, while others make the opposite choice, preferring material wealth and rapid income growth; there is no general reason why the latter should redistribute their wealth to the former. Certainly, any difference in wealth or the satisfaction of human needs worldwide is not attributable to the free choice of values by citizens of different states, but Rawls did not intend to reflect directly on the empirical condition of the contemporary world. His purpose was to determine the relationships to be established between viable states, meaning self-supporting and stable polities. This implies that the states considered would ensure the satisfaction of the basic needs of their population. Where, empirically, some states cannot meet these viability conditions, Rawls noted the duty of assistance to failed states, to help those in need to set up the required institutions.

Another argument reinforced in Rawls's thinking the asymmetry in treatment between 'domestic' justice, internal to a state, and 'international justice'. For domestic justice, the people who deliberate under a veil of ignorance nevertheless know they are in the process of defining the rules of a society that will have the necessary institutions for the establishment of the rule of law, including in terms of the use of coercion. To the contrary, when state representatives are supposed to meet to define the rules of international relationships, they know there is no such thing – at present or in the future – as an authoritative world government, such as one that ensures domestic justice. Relationships of mutual trust in the behaviour of others, which depend on the effectiveness of the law, can only be weakened or lacking at the international level, limiting the level of commitment to others.

The debate raised by the 'Law of Peoples' enables the identification of two important and controversial points generated by any attempt to extend justice requirements from domestic to international relations: (1) how conceptions of justice tackle the pluralism of the conceptions of the good to be pursued; and (2) how the principles of justice depend on the institutional context bordering the deliberation of a constituent assembly.

3.2.3 The Pluralist Internationalism of Mathias Risse

Mathias Risse (2012) aimed to go beyond the radical alternative between authors who limit the relevance of the idea of distributive justice within the national space, and those like Charles Beitz (2005) or Richard Arneson (2005) who apply this distributive justice to all people indiscriminately. For the latter, the boundaries that separate and distinguish states are contingent and morally arbitrary elements. Reviving the viewpoint of Hugo Grotius, Risse adopts as a starting point the idea that justice materializes in different ways that echo distinct foundations. He considers various foundations: common humanity, collective ownership of the Earth by humankind, participation in the geopolitical order and participation in the global trading system. Thus, each foundation generates specific obligations. While recognizing the importance of the special relationship established between citizens of the same state, Risse therefore believes that they also have obligations in the name of other foundations to those who are not their fellow citizens.

Common humanity provides the foundations for the natural rights of all human beings, who by birth, are morally entitled to the means to survive and to live a decent life. Collective ownership of the Earth gives legitimacy to the demand for a fair distribution of natural resources and territories, so that everyone can meet their basic needs. Links created by global trade raise the issue of the fair distribution of the gains from trade and respect for the rights of trade partners. For Risse, in a world that has become 'globalized', these obligations are also those of states and international organizations, since individuals and states are engaged in forms of cooperation and coercion at this international level, as illustrated by the World Trade Organization.

Nussbaum (2013, p. 481) endorsed such an approach of non-exclusive rights and duties that have different foundations and give rise to a multi-layered system of norms:

> People, then, value their own nation intrinsically and non-derivatively, but they also value all human beings, and seek a world in which all are raised up above a threshold of capability . . . We may love our country most, and even seek more prosperity for it than for others, so long as we also seek a reasonable threshold level of capability for all.

Consequently, international redistribution, if any, should respond to specific norms that are different from domestic ones, and find means of modulation capable of respecting national sovereignty and the specific balance established by each state between the various common goods pursued.

3.2.4 The New Cosmopolitan Frontier

A radical moral positioning

Since the 1990s, an ancient stream of analysis – cosmopolitanism – has re-emerged. Dating back to the Roman Stoics, its starting point is the idea that there is a universal moral order that everyone can access by means of rational deliberation. The outcome is a set of principles, rules and duties that every 'citizen of the world' should take on to ensure universal respect for the fundamental rights of every human being. The empirical source of the contemporary revival of this movement lies in the widespread fabric of social and economic interdependencies throughout the world. In the eyes of cosmopolitanist authors, the interdependence of the world is such that humankind, objectively, has already evolved into one single polity.

From an analytical perspective, the contemporary foundations of this approach are threefold:

1. Pursuant to ethical individualism, only the individual human being is a moral subject and a moral agent and, as such, the focus of attention of moral reflection; collective entities, such as family, clan, nation or network, are only derivative constructions, the value of which must be measured by their contribution to the moral achievement of individuals.
2. From the point of view of fundamental rights and the moral status of individuals, distinctions and separations between groups or communities, particularly among nation states, are arbitrary and contingent: to be born in one society rather than another, at one date rather than another, are contingent facts, as are the borders and other boundaries that should not interfere with the status, rights, consideration and justice to which individuals are entitled worldwide.

3. The quality of being human gives every person equal moral worth and dignity; consequently, one should have very strong reasons for exceptionally waiving the principle of equality of all regarding access to the goods valued by individuals.

Constructed in this way, moral cosmopolitanism is opposed to various forms of particularism or communitarianism, whatever their manifestations. For some authors (Held, 2005; Renaut, 2013), these moral roots lead to a form of political and institutional cosmopolitanism. They try to characterize the new institutions that should be set up globally to give full effect to the cosmopolitanist principles.

Meanwhile, this issue of global institutions is a source of division among cosmopolitanist authors. Some (Pogge, 1994) propose taking Rawls's method against the Rawlsian precepts, namely scheduling what would be a rational deliberation of the 'citizens of the world', potentially belonging to all possible generations, placed under the veil of ignorance, to deliver distributive principles such as those that Rawls identified for the internal justice of a state. Others (Chauvier, 2006) believe that the arguments and recommendations developed at the level of the nation state cannot be extended to the global level if they remain unchanged: global justice is not domestic justice at the global level. Cosmopolitanist thinking has therefore focused on several areas of development, as described below.

The content of redistributive duties between states For those who admit the existence of an international redistribution requirement, the issue is to determine what variable it should be based upon. An old controversy has been revived between those who primarily want transfers of resources (finance, technology) to restore equality, and those, for example Simon Caney (2011, pp. 90–97) in the case of climate change, who propose focusing thinking and action on 'needs' and 'capabilities', over and above resources. For Alain Renaut (2013), however, it is the well-being of people that should be the target, as this multidimensional concept integrates both 'resourcist' and 'capabilist' approaches by encompassing both the material means of developing life-projects and fundamental liberties.

The moral and political foundation of global redistribution To postulate the need for global redistribution is one thing; to give it a clear foundation, which would simultaneously provide the motivation for its worldwide implementation, is another. Chauvier (2012, p. 7) calls for a reflective equilibrium between normative intuitions and reasons on the one hand, and pragmatic social engineering on the other. Nussbaum (2013, p. 486) goes in the same direction when she recalls that 'a norm should be non-utopian but aspirational'; although Sen (2009) has expressed his doubts about normative efforts to shape comprehensive aspirational theories, and has urged philosophers of justice to determine how to reduce injustices in the real world. In this regard, several authors have underlined the fragility of a global redistributive justice based only on the moral adhesion of individuals to the rightness of the cosmopolitanist viewpoint. It is necessary to take account of the motivations and interests of the richest societies, in which redistribution would originate. The fragility of a regime depending only on each individual's voluntary adhesion to the moral principles of cosmopolitanism appears so great that Renaut (2013) believes it inevitable that redistribution will be founded for the greater part on 'the well-understood best interests' of states. He thus meets one of the conditions identified by Rawls (1971 [1999], pp. 109ff.) concerning the relevance of the political conception of justice. As already noted, the latter is constrained by the conditions of cooperation between agents who are both rational and reasonable, and mutually disinterested, being neither altruists nor saints. Individuals know they need the cooperation of others to be able to achieve their own goals. To this effect, they should take account of the existence and intentions of others in order to obtain their cooperation.

Criticism of the injustice of the economic structures and international institutions in place The extreme levels of poverty and inequality worldwide are said to be caused by the structural injustice of the international institutions. This unjust state prevails because the latter maintain the peripheral and intermediate countries in a state of dependence on the 'core' of world capitalism and impose ecologically unequal trade (Martinez-Alier, 2002; Rice, 2007; Roberts and Parks, 2007). The ultimate responsibility is placed on the shoulders of the inhabitants of rich countries who continue to support these unjust institutions and structures; for this reason, they have a moral obligation to compensate the victims of these injustices (Pogge, 1994).

The arbitrary and unfair state of natural resources ownership The issue of the appropriation of natural resources is a long-standing subject of debate among philosophers and economists, which has not produced any closure or definitive answers. According to cosmopolitanists, states should not have full ownership of the natural resources in their territory (Chauvier, 2010; Risse, 2012) and should not be allowed to appropriate common natural resources over and above their fair share. Present ownership rights, says Beitz (1979), are one of the major causes of inequalities in the world. Thus, Thomas Pogge (1994) proposed a tax of 1 per cent on the annual value of natural resources consumption globally, to finance poverty eradication. Various suggestions have emerged along similar lines (Hayward, 2005).

The lack of democratization in international institutions and networks Whereas tradition associates democracy with local and national governmental structures, international centres of economic and political power reveal large unaccountable hierarchies of influence and vast inequalities of access and control by citizens. David Held (2005) underlined the need for a clear distinction between the democratic idea and the nation state ideal. The nation state is neither the only nor the main vehicle for affirming a community. It is not the one and only political space in which people can self-determine their collective rules. Nowadays, identities and allegiances are multiple and overlapping.

The search for a balance between fairness towards individuals worldwide and the sovereign freedom of political communities Some cosmopolitanists consider the distinction between hundreds of nation states as an arbitrary element that should be set aside from any moral argumentation about justice. Other scholars try to develop principles that can unify acknowledgement of the intrinsic value of a population's capacity to self-determine social values and rules, paving the way for the emergence of political pluralism, and universal norms of justice for individuals. Refusing to take the existence of nation states as an evident and given starting point, Chauvier (1999) questioned the conditions for a fair division of the presumed unified human community into different political societies. How could this separation be more valuable than maintaining humankind within a single political society? Which features make the subsequent political pluralism intrinsically valuable? Exploring various combinations of social values, and considering freedom of choice as an intrinsic value, provide some answers, but there should be counterparts from the viewpoint of Rawlsian individuals. The author imagines a debate on this issue of separation under a veil of ignorance, and concludes that an assembly of individual representatives of humankind would wish to establish a set of safeguards:

- One should plan for compensation to offset the inevitable inequalities of natural resource endowments, since the composition of the Earth is heterogeneous.
- One should establish procedures for the ongoing correction of historical drifts, by providing, via trade, guaranteed access for all to essential natural resources for economic development; states should be prohibited from withholding critical resources located in their territory

- Individuals should have the right to leave the state in which they were born to join another.
- Well-ordered states should have a duty to intervene in the affairs of states that are significantly disturbed, to ensure the respect of individual rights.

Undoubtedly, these principles go beyond the current principles governing the organization of the international society, but also the formalization of Rawls's Law of Peoples. They tend to give certain cosmopolitan rights to individuals without denying the intrinsic value of an international society organized as a society of nation states. Note that a notion of distributive justice is not directed at wealth or income, but is limited to the field of natural resources (ownership, access and compensation).

The acknowledgement of certain goods as being 'of cosmopolitan interest' (Chauvier, 2010) Cosmopolitan goods require specific norms of devolution and use. They are those that concern all people worldwide to the same degree. The author presents two justifications for this conceptual innovation: either these goods are natural and, being essential, are common goods to all human beings; or, being unnatural, they have a universal human value. This concept converges with the view of Risse, who built his proposals on the assumption of the initial common ownership of the Earth by humankind. For both authors, this does not mean that every piece of land should be maintained under collective ownership and management by the whole of humankind. It asks for the procedures used for their allocation to consider the interests of all humankind, which include both present and future generations. For this class of goods, states should act as the agents of humankind from a top-down approach, including duties to report to the community of states on the way these resources have been managed and used.

Would it be possible to implement all of these ideas without having a true world government? Chauvier warns that we should not expect too much in practice from cosmopolitanism. Analyses inspired by it willingly assume that democratization can be achieved on any scale. Such a judgement is disputed, at least since Aristotle. As the geographical and political scale expands, so do the problems raised in terms of reaching agreements on shared conceptions of the common good or the sense of justice. Any proposal envisaging the significant, if not massive, worldwide redistribution of wealth will encounter these difficulties. There is no doubt, for Chauvier, that democracy, justice and redistribution can only fully develop on scales that foster mutual acquaintance, direct links and social trust.

Between a global egalitarian moral order and the conditions for international cooperation

The proposed survey reveals a tension between moral and political conceptions of cosmopolitan justice. Put simply, for the purely moral approach, the word 'justice' is a vehicle for affirming the value of equality in all areas of social life. This pre-eminence is said to be the necessary consequence of the premise of the equal moral dignity of all human beings. This goes very far. It will not be enough to ensure equal opportunities for all: equal possessions are also required, as well as equal welfare for all, unless there are good reasons for exceptions to the egalitarian rule.

Yet as Miller (2007, pp. 24–31) and Jocelyne Couture (2010, pp. 17–20) have shown, the initial assumption of the moral equality of all human beings does not imply that all people have the same general obligations to each other. Nor does it settle the question of what to equalize. Take the example chosen by Miller: everyone agrees to grant the same moral status to every child born; it does not follow that a father has the same duty of care to all children in the world. One pioneer of modern thinking on cosmopolitanism, Nussbaum (2008, p. 311), clearly explained the point:

> Politics, like child care, will be poorly done if each thinks herself equally responsible for all, rather than giving the immediate surroundings special attention and care. To give one's own sphere special care is justifiable in universalist terms . . . we do not really think our own children are morally more important than other people's children, even though almost all of us who have children would give our own children far more love and care than we give others'. It is good for children, on the whole, that things work this way, and that is why our special care is good, rather than selfish.

Therefore, it is from a cosmopolitanist point of view that it is a good thing that every human being does not act equally towards all others. This point again raises the question of which link to establish between the fundamental value of the moral equality of all human beings across the world and the specification of the moral duties of every person regarding all others. In line with Rawls, Couture (2010, p. 19) called for cosmopolitanist individual egalitarianism to be replaced with the duty of assistance to peoples in the development of just institutions that conform to their sense of justice. International justice would have 'a moral obligation to create, consolidate or at least not to oppose the conditions that foster the sense of belonging, cohesion and cooperation within each people'. This rewording does not support formal equality between individuals, but 'establishes equality among peoples in their ability and right to self-determination as well as the absolute right of citizens to take part in the shaping of their own society' (ibid., p. 35).

With this confrontation, the idea of global justice is divided between a worldwide egalitarian moral order and a political conception that gives prominence to the self-determination of peoples. The first encloses individual behaviour in a dense network of obligations towards all 'citizens of the world'. The second highlights the conditions for the emergence and self-determination of well-ordered sovereign societies, in which freedom of democratic expression and deliberation would ensure that autonomy in the pursuit of the good and the just is not empty words.

Cosmopolitanism tends to see itself as a principle for the total organization of the global society, leaving purely instrumental roles to the political and legal orders, particularly on the scale of nation states. By considering the existence of political communities and other forms of association as arbitrary facts that are irrelevant to moral reflection, it also actively works to delegitimize the institutions of political communities and the values they embody.

In line with Rawls, the second approach reveals a more modest ambition. Equality and justice are not the first terms of the reasoning, but the conditions for cooperation between human beings. From this point of view, the differences built between internal justice in a political community that has developed state institutions, and justice in the relations between states, reflects major differences in the conditions for human cooperation.

This second approach requires an adapted moral theory that incorporates an element of indeterminacy to uphold the legitimate margin of autonomy for peoples to express their sense of justice in the definition and ongoing adjustment of the social contract governing their collective life. The refusal of moral perfectionism that would subject the political organization to a specific moral order leads to recognition of the pluralism of the common goods that may be sought within each society and between states, but also the pluralism of conceptions of the just, which is perfectly reflected by debates between scholars on the issue of global justice.

Where moral and political universes meet, various scholars such as Nussbaum, Chauvier and Renaut have found an intermediate place. They are close enough to cosmopolitanism, but also sensitive to the democratic values expressed in political communities. They wonder about the subjective and political conditions for the incorporation of some cosmopolitanist elements into the reality of international relations. Others, like Miller, reject any cosmopolitanism, but refuse to abandon international relations to pure realism or the cynicism of the balance of power. In their final proposals for combinations of rights and duties of different types and foundations, they are not so different from one another.

NOTES

1. Some powerful stakeholders have developed a strategy to disseminate doubt about the soundness of climate science in the public opinion in order to weaken popular backing for climate policies. On who these stakeholders are and how they operate, see Michaels (2008), Hoggan and Littlemore (2009) and Oreskes and Conway (2010).
2. 'Climategate' resulted from the internet release of thousands of hacked emails from climatologists linked with the University of East Anglia, UK team of scientists who were dedicated to studying the global climate by producing and gathering numerous data on temperature levels across the world. Some extracts from these emails were interpreted by climate-change sceptics as revealing scandalous efforts aimed at: (a) hiding manipulation of temperature data to align them with the 'dogma' of global warming; and (b) trying to coordinate a network of climatologists around the goal of blocking access of papers critical of the 'dogma' to scientific journals controlled by mainstream climatologists. Several enquiry commissions have confirmed the integrity of practices by Jones and his colleagues, but have asked for greater efforts towards the transparency of scientific data.
3. One referred to the date at which all Himalayan glaciers would melt, given as 2035 (which has no scientific basis whatsoever) instead of 2350; the second concerns the proportion of the Netherlands that could be flooded by seawater – this was in fact the proportion of the country susceptible to flooding either from the sea or from rivers.
4. In a non-cooperative equilibrium, each party optimizes its behaviour as a function of its own interests (equalization of individual marginal costs and benefits), assuming the behaviour of others as given and based on a common attitude of 'each for himself'.
5. In such games, there is no net benefit from staying outside.
6. In this game, two players must decide how to divide a sum of money between them. The rule is that one player makes a proposal and the second either accepts or rejects it. If accepted, the proposal is implemented. If rejected, none of players receive any money. This is a one-shot game. Experience shows that proposals giving less than 20 per cent of the sum to the second player are often rejected.
7. Acknowledged in the 1980s, the concept of the 'common heritage of humankind' came up against political difficulties during the negotiations for the Convention on Biological Diversity that was eventually adopted in June 1992. This convention finally settled on the principle of states' sovereignty over their biological resources rather than that of the 'common heritage' initially supported. In defence of the latter concept, see Brown-Weiss (1989).
8. In the case of climate change, see for instance the pioneering work of Agarwal and Narain (1991), Grubb (1995), Kverndokk (1995), Gosseries (2004a) and Shue (1999).

4. Justice and climate change: data and proposals

In what follows, I assume that the main features of the scientific background to climate change are common knowledge to readers. Box 4.1 gives some definitions and notations. I would, however, like to underline some features that are critical to the discussion about climate justice.

BOX 4.1 DEFINITIONS AND NOTATIONS

Radiative forcing refers to changes in the Earth's radiation balance. It is expressed in watts per m^2. Anthropogenic GHG emissions affect the climate via radiative forcing. Changes in solar activity and volcanism are other forcing factors.

The main greenhouse gases contributing to climate change are carbon dioxide (CO_2), methane (CH_4), nitrous oxide (N_2O) and various fluorinated compounds characterized as ozone-depleting substances (ODS).

CO_2 is the most important of the GHGs due to its atmospheric lifetime explained by the balance between atmospheric intake and outtake by oceans and terrestrial ecosystems; after 100 years, about 30 per cent of the CO_2 initially injected into the atmosphere remains (Hansen et al., 2013), whereas the whole elimination process may take several millennia. CO_2 is used as a unit of account for other GHGs, where each gas is apprehended based on its impact on radiative forcing. We then move from CO_2 to CO_{2eq} to designate a 'CO_2 equivalent'. Quantitative emissions evaluations can be given in tons of carbon (tC), tons of CO_2 (tCO_2), or tons of CO_{2eq} (tCO_{2eq}); 1 tC is equivalent to 3.667 tCO_2. Millions of tons are written Mt, and billions of tons Gt.

Ppmv stands for 'parts per million by volume'. In this book, I will use the shorter 'ppm' for the same definition; 1 ppm equals a millionth, or 10^{-6}; 1 ppm of CO_2 stored in the atmosphere is equivalent to 2124 GtC or 7.788 $GtCO_2$. Given an average atmospheric emissions retention rate of 45 per cent, observed during the second half of the twentieth century (Ballantyne et al., 2012), it will take about 17 $GtCO_2$ emissions to increase the atmospheric CO_2 concentration by 1 ppm in the coming decades. But it is assumed that the retention rate is an increasing function of atmospheric GHG concentration. For a 1000 ppm concentration, the rate could grow to 55 per cent. This means that to obtain 1 ppm more in the atmosphere, gross emissions should be 17 $GtCO_2$ with a 45 per cent retention rate and 14 Gt with a 55 per cent rate.

4.1 SCIENTIFIC AND GEOPOLITICAL FEATURES OF THE CLIMATE THREAT

4.1.1 Of the Morphology of the Climate Issue

Climate disruption is not directly attributable to greenhouse gas (GHG) emissions fluxes, but rather to the stock of gases accumulating in the atmosphere and the carbon and heat stored in the oceans. Several works (Allen et al., 2009; Matthews et al., 2012) have proposed reasoning in terms of a budget for cumulative emissions over a long period, such as a 'carbon budget' from 2010 to 2050, without focusing solely on an emissions target for a given date such as 2050. This proposal converges with the idea of recording cumulative GHG emissions over a long historical period with the purpose of assigning responsibility for the climate threat between countries according to their historical contributions.

Global risks due to climate change are mainly independent of the geography of GHG emissions, but their impacts will affect different regions in very different ways, despite the instability of this distribution regarding temperatures, rainfall and extreme events. The global climate machine makes it impossible to identify a link between given emissions from one place and given impacts in another. The unequal distribution of impacts has no link with the geographical distribution of most GHG emissions.[1]

Climate change is surrounded by all-pervasive uncertainty due to the global, very long-term and dynamically complex nature of the issue, but also to historical uncertainty regarding future GHG emissions trajectories in the twenty-first century. The general concept applying to such a context is that of risk, in its broader meaning, which covers uncertainty. Present global climate change can almost certainly be linked to GHG emissions since the nineteenth century (IPCC, 2007), and especially since 1990. But it is and will be impossible to establish a direct causal link between identified GHG emissions and the identified, localized damage contributed by specific weather events. Every time an extreme event occurs, scientists explain that it is impossible to assert a direct link with global climate change. We can only speculate on statistical presumptions about overall tendencies, not on individual events. As everyone knows, there is still scientific and political debate about the respective roles of natural causes and human responsibility in creating the current climate change, and still more about the explanation of specific events. This certainly puts some major obstacles in the path of founding climate justice on liability for specific 'damage' caused to specific 'victims' by specific 'polluters'.

4.1.2 From Doubt to Change of Approach

After the diplomatic success of the Kyoto Protocol and the refusal of the USA to join it, a period marked by doubt, hesitation, exploration, failure and tentative recovery began. The divide between two regimes of action, one for the developed countries and another for the emerging and developing countries, has become unsustainable for most industrialized countries in discussions about how international action should be organized in the post-2012 period, whereas the developing countries have clung to this divide at all costs. The premise of a solution overcoming this divide was introduced during the Copenhagen Conference of the Parties (COP) in December 2009. This COP was the turning point in terms of approaches, switching from a top-down rationale, which has been tested since the Kyoto Protocol, to a bottom-up approach. This has had one effect: placing all countries within the same general framework, without abandoning the three criteria of justice set out in the United Nations Framework Convention on Climate Change (UNFCCC): equity, common but differentiated responsibility and capabilities. Uncertainty as to the future existence of international cooperation to prevent climate change was eventually resolved, on paper, with the conclusion of the Paris Agreement in December 2015, to be confirmed by actual concerted action to come in the future.

This agreement confirms the change of approach initiated at the Copenhagen Conference: instead of seeking an agreement on criteria for allocating a carbon budget to each developed country, the Paris Agreement sets a long-term goal – keeping the global temperature increase in the lower atmosphere well below 2°C, and as near to 1.5°C as possible – and asks all countries, whether developed or developing, to communicate their own mitigation targets. Surprisingly this agreement acknowledges that the aggregate mitigation efforts that would result from announced national targets from all countries, following the bottom-up approach, fall far short of meeting this goal, and asserts the need for future political dynamics to reduce this gap.

4.1.3 Evolution of GHG Emissions and Concentrations

With the passing of time between 1990 and 2015, the objective situation, including the respective weight of GHG emissions from each region of the world, has changed radically since the adoption of the UNFCCC. China's annual emissions tripled between 1990 and 2012 and have increased more than tenfold since 1970, making this country the world's largest emitter, accounting for 25 per cent of total emissions in 2012; other countries such as South Korea and Saudi Arabia more than doubled their emissions over the same period of 1990–2012. Meanwhile France, for instance, reduced its emissions by 12 per cent over that period.

Since 1850, atmospheric GHG concentrations have increased dramatically. For CO_2, the atmospheric concentration increased from 280 ppm in 1850 to almost 390 in 2010, a level not seen for hundreds of thousands of years. Between 1990 and 2010, every year brought an increase of between 1.8 and 3 ppm of CO_{2eq} in the atmosphere. The level reached in 2013 by all GHGs amounted to 479 ppm CO_{2eq}, representing a 34 per cent increase in radiative forcing from 1990 levels (World Meteorological Organization, 2014, p. 3). Consequently, the average temperature of the lower atmosphere has increased by about 0.8 °C since 1850.

4.1.4 The Data on Countries' Historical Emissions

It was Brazil that first proposed in 1997 that GHG emissions reductions by the developed countries should be in proportion to the radiative forcing generated by their historical emissions since the beginning of the industrial era (Pinguelli-Rosa et al., 2003). Although the idea of accounting for all GHG emissions over a century and a half may sound simple, the statistical data proved more difficult to obtain for the earliest years, for the developing countries, for deforestation and other changes in land use (urbanization, cultivation, and so on), and for non-CO_2 GHGs such as methane from rice paddies. Some emerging and developing countries (China, Brazil, India, and so on) and non-governmental organizations (NGOs) took advantage of difficulties and uncertainty to put forward estimates based solely on CO_2 emissions related to the use of fossil fuels for power generation, cement production, industry, transport and housing (heating). Others omitted to consider the major recent trends observed since 1990. In contrast, the developed countries have worked to ensure full accounting of all GHG sources, incorporating the most recent data.

To investigate the statistical issue underlying the political claim concerning the historical responsibility of the North, a working group was set up in 2003 as part of the UNFCCC Subsidiary Body for Scientific and Technological Advice (SBSTA). Its report (Höhne et al., 2008) resulted in the following distribution of historical contributions of the different regions of the world until 2000: 40 per cent for the Organisation for Economic Co-operation and Development group of countries as of 1990 (the OECD 90), 14 per cent for Eastern European countries and the former Soviet Union, 24 per cent for Asia, and 22 per cent for Africa, Latin America and the Middle East.

Amendments were proposed from different sides. Some wanted to subtract emissions corresponding to the satisfaction of 'basic needs' (food, energy, and so on) from developing countries' accounts. Depending on the proposals, this figure stands at between 4 and 7 $GtCO_{2eq}$ of annual emissions that should be deducted globally. In line with the goal set out in the Paris Agreement, humankind should move from an emissions level of 30 $GtCO_2$ from energy sources in 2020 to almost zero in 2050 in order to stand a reasonable chance of remaining below the 2°C threshold. According to advocates, 'survival' or 'essential' emissions should be allocated on an equal basis to the inhabitants of each country (Müller et al., 2009).

Others have introduced restrictions on the scope of considered emissions. For instance, a Ting Wei et al. (2012) study funded by the Chinese government excluded emissions caused by massive deforestation in many developing countries in the second half of the twentieth century, since a substantial amount of the timber harvested was destined for developed country markets, in countries said to have already exhausted their own forest resources.

The most comprehensive assessment available is by the team of Michel den Elzen et al. (2013) at the Netherlands Environmental Agency. It considers the evolution of historical emissions until 2010. Some results are reported in Table 4.1. Depending on the gas and the sources integrated, the picture given of emissions from the different regions of the world varies considerably. By integrating all GHGs for the 1850–2010 period, the relative share of cumulative emissions reaches 52 per cent for the developed countries and 48 per cent for the emerging and developing countries. Parity of cumulative emissions between the North and the South was reached around year 2015 and the proportion will reverse by 2020. If only recent emissions are considered – those produced since 1990 – then the developed countries are accountable for 43.5 per cent of total emissions and the South for 56.5 per cent. When the calculations include only CO_2 from energy sources, the historical contribution of the developed countries reaches 71.4 per cent of the total, while the emerging and developing countries see their account reduced to 28.6 per cent.

The developing countries insisted that data should be presented by calculating the per capita emissions of each country, instead of total emissions per country. This request is usually justified by the idea that every 'citizen of the world' has the same rights regarding the capacity of the atmosphere and other natural systems (oceans, terrestrial ecosystems) to absorb GHGs. Due to demographic factors, this presentation particularly reduces the causal responsibility of the South and symmetrically increases that of the North. By dividing cumulative emissions since 1850 by the cumulative population of each country over the same period, the mean

Table 4.1 Contributions of countries to cumulative GHG emissions

Countries	Cumulative emissions (% of total)		
	All GHGs 1850–2010	CO_2 only, excluding emissions linked to land use (deforestation, etc.) 1850–2010	All GHGs 1990–2010
United States	18.6	27.6	15.6
EU-27	17.1	24.8	12.4
Russia	7.2	8.0	6.5
Japan	2.8	4.2	3.3
Australia and New Zealand	1.7	1.3	1.7
Canada	1.9	2.1	1.8
Other European countries	1.1	1.4	1.0
Ukraine	1.5	2.1	1.3
China	11.6	10.5	15.3
India	4.1	2.7	4.5
Brazil	3.9	0.9	4.1
Indonesia	4.8	0.6	6.2
South Korea	0.6	1.0	1.2
Mexico	1.3	1.1	1.6
Africa	7.1	2.4	7.5
Total developed countries	51.9	71.4	43.5
Total emerging and developing countries	48.1	28.6	56.5

Notes: This table underlines the contrast between historical contributions of countries depending on the choices made for the scope of GHGs and the period. The percentages of the US or the European Union increase by 50 per cent when switching from an all-GHG base to one including only CO_2 from fossil fuels. Note that India's contribution reaches between a quarter and a third of that of China, that Chinese and US cumulative emissions are similar over the 1990–2010 period, and that Africa's contribution since 1850 is similar to that of Russia. Overall, the picture undermines the commonly held belief that the developed industrial countries have almost exclusive responsibility for the climate problem, while the emerging and developing countries form a group that includes only victims.

Source: den Elzen et al. (2013, pp. 402–3).

per capita contribution of the developing countries is 70 per cent of the average rate, while the per capita contribution of the developed countries is 1.9 times the average rate (den Elzen et al., 2013). With this method, the historical Northerner produced 2.7 times more GHGs than the historical Southerner.

Caution is advised, however, when interpreting these results. Looking at per capita emissions percentages can rapidly lead to major biases of reasoning for insufficiently attentive readers. Consider the fictitious case of a world in which a country X with a population of 1 million inhabitants annually emits 1 $GtCO_{2eq}$ out of a global total of 50 Gt. The inhabitants of other countries, amounting to 7.2 billion, together emit 49 $GtCO_{2eq}$. Emissions per capita in country X are 1000 tCO_{2eq} and emissions from other countries are 6.8 tCO_{2eq}. The average inhabitant of country X thus emits 147 times more gas than the average inhabitant of the other countries. Yet it cannot be claimed that he must assume 99 per cent of the financing required to compensate for the other countries' losses and to ensure their adaptation to the new climate. Despite the high level of per capita emissions, his country accounts for only 2 per cent of global emissions.

Finally, some authors have proposed going beyond the issue of accounts per country by calculating the carbon footprints of consumers in different countries. With these calculations, consumers are charged not only for their direct emissions, but also for attributed indirect emissions. The latter come from the national territory due to various input activities (production, transport) for the goods and services they consume, and from the production process and transport of goods imported from abroad. For example, in 2007 the French domestic carbon footprint of consumers was estimated at 750 $MtCO_{2eq}$, while emissions from the territory were only 506 $MtCO_{2eq}$. The contrast in representations is important: while emissions from the territory fell by 7 per cent between 1990 and 2007, the carbon footprint of overall French domestic demand was up 14 per cent over the same period (CGDD, 2012).[2] Indeed, during this period the French economy further entered the global economy, and some of the most GHG-emitting production activities were transferred to other countries, particularly China or other Asian countries. Worldwide, from 1990 to 2008, international trade transferred about 16 $GtCO_2$ from the developed countries to the developing countries, mainly China (Peters et al., 2011).

Shaped by the intention of allocating responsibilities and assigning future rights to countries, the accounting of historical GHG emissions is the persistent subject of contrasting presentations that have failed to converge on a shared vision of historical data or of the role to give them when defining the rules for the future. These presentations of data incorporate the moral and political theories of those who advance them. Some methodologies are dubious. In most cases, choices of methods depend on assumptions that should be subject to moral criticism. This is the case of the aggregation of cumulative emissions since 1850 to assess countries' responsibility, of calculations of consumers' carbon footprints, or of presentations imposing per capita values as a self-evident basis for any allocation of rights and obligations.

4.2 MAIN PROPOSALS FOR EQUITABLE BURDEN-SHARING

Since 1992, many analyses and proposals have been made in the context of international negotiations to devise just and fair principles to govern burden-sharing for mitigation and adaptation policies or to allocate GHG emissions rights to countries. These proposals have come from different sides: governments, NGOs, scholars, scientists and experts of all sorts, very often with overlaps and synergies between these groups. I outline some representative proposals below.

4.2.1 Governments

Bolivia

The developed countries should compensate the developing world for the damage already done and for obstacles to their future development imposed by constraints on the use of fossil energy sources (State of Bolivia, 2009).

Justification Victims should be compensated for any kind of damage done to them, in order to restore equality of development opportunities.

Brazil

For the developed countries – and only them since the developing countries should be exempted from any quantitative restrictions inasmuch as they have not reached the development level of the developed countries (Pinguelli-Rosa et al., 2003) – obligations to cut emissions should be based on countries' historical responsibility for inducing global warming. In practical terms, this requirement has been translated as commanding a distribution of rights that is inversely proportional to cumulative GHG emissions since the beginning of the industrial era (1850).

Justification Polluters should pay.

France

The allocation of GHG emissions targets to industrialized countries should involve a process of progressive convergence of long-term emissions rights based on appropriate indicators (population, abatement cost, GDP) (French government, 1997). Specific financial resources should be directed to less-developed countries for adaptation and support for the adoption of low carbon technologies.

Justification Long-term economic equality of individuals is to be ensured in a world where development inequalities should progressively disappear.

India

The distribution of emissions rights to countries should be based on their respective populations (Agarwal and Narain, 1991; Government of India, 2009).

Justification The atmosphere is a global public good; every citizen of the Earth has an equal right to use the atmosphere.

USA

All countries should commit to controlling their emissions according to goals and targets they set themselves (bottom-up approach), while considering the commitments of other countries (the US would never join a regime in which the emerging countries do not make significant commitments).

Justification States are sovereign and no international institution should impose anything on them: a climate change regime could only result from the voluntary cooperation of states based on a realistic account of what each can do. Since the 2009 COP in Copenhagen, this approach has generated the 'pledge and review' regime that had been advocated, in some cases for a long time, by well-known scholars such as Schelling (1992) and David Victor (2011).

4.2.2 NGOs

By the 1990s, civil society, with both environmentalist and pro-development NGOs, was extremely present around international negotiations, albeit with tensions between them. Despite the early emergence of new concepts such as 'unequal ecological exchange' and the 'ecological debt' of the North, it was not until 2000 and the Hague Conference that militant groups and advocate scientists from environmentalist circles and grassroots organizations joined in speaking on behalf of local communities and pooled their networking efforts to influence the course of negotiations in the name of 'climate justice'. Among their claims was the requirement that the North repay its 'ecological debt' to the South, and a harsh denunciation of the Kyoto Protocol, which was accused of exclusively serving the interests of the North and multinationals (Roberts and Parks, 2009, p. 394). Subsequently, a huge NGO coalition was formed in 2002 as the International Climate Justice Network. This network includes CorpWatch, Friends of the

Earth International, Greenpeace International, Groundwork, Indigenous Environmental Network, Indigenous Information Network, National Alliance of People's Movements, National Fishworkers Forum, OilWatch Africa, OilWatch International, Southwest Network for Environmental and Economic Justice, Third World Network and World Rainforest Movement. Meeting in Bali in August 2002, the coalition drafted the *Bali Principles of Climate Justice* (International Climate Justice Network, 2002). The main elements are the following, with a comment by myself extracting the underlying normative idea:

1. Communities have the right to be free from climate change, its related impacts and other forms of ecological destruction. The result is a general obligation for all countries to reduce their emissions with a view to completely eliminating them in the future.
   Underlying idea: the right of peoples to a stable climate.
2. Affected communities and indigenous peoples have the right to be represented in negotiations and to speak for themselves there.
   Underlying idea: justice requires equitable access for all to the negotiation process, which should first benefit the local communities.
3. Climate justice opposes the role of transnational corporations in shaping unsustainable production and consumption patterns and lifestyles, as well as their role in unduly influencing national and international decision-making.
   Underlying idea: ecological sustainability, understood as the maintenance of local ecosystems and biodiversity, is posited as a principle of justice for the benefit of communities and indigenous peoples.
4. Climate justice calls for the recognition of a principle of ecological debt that industrialized governments and transnational corporations owe the rest of the world because of their appropriation of the planet's capacity to absorb GHGs; victims of climate change should receive full compensation, restoration and reparation for loss of land, livelihood and other damages.
   Underlying idea: recognition of the full 'historical responsibility' of developed countries to compensate and repair.
5. All people, including the poor, women, rural and indigenous peoples, have the right to have access to affordable and sustainable energy, in other words renewable energy.
   Underlying idea: organizing access for all communities to clean energy technologies is a requirement of justice.
6. Climate justice calls for a moratorium on all new exploration of fossil fuel deposits, but also on the construction of large hydroelectric dams and nuclear power plants.

Underlying idea: all technological developments that are damaging to the climate and the environment of communities should be banned.

7. Climate justice affirms the rights of communities that are dependent on natural resources for their livelihood and cultures to own and manage these resources in a sustainable manner, and is opposed to the commodification of nature and its resources.
   Underlying idea: the hubris of the market is inherently adversarial to genuine community interests, since it imposes the degradation of ecosystems and natural resources.

These Bali Principles certainly deserve a critical examination, which will be done later in this book, especially for principles 1, 2, 4 and 7. They were followed in 2004 in the United States by a similar 'Climate Justice Statement' drafted by a group of researchers and activists. As an outcome of its strategy of influence, the network was able to penetrate certain national government circles – in Germany and the United Kingdom, for instance – and the top management of international organizations such as the United Nations Development Programme (UNDP). In 2009, the Third World Network addressed a request to the government representatives meeting in Copenhagen entitled 'Repay the climate debt!', which was signed by 254 organizations (Warlenius et al., 2015b, p. 23), without success.

The mobilization and activism of NGO networks around the climate negotiations for more than a decade have reinforced the perceptions and normative beliefs of most developing countries on certain ideas: (1) the problems affecting the environment across the Earth are mainly attributable to production and consumption patterns in the North; (2) the ability to act on these issues depends primarily on the position of countries in the international division of labour (centre or periphery): the centre holds the power, while the periphery can only suffer and follow; and (3) the Northern governments are manipulating environmental issues to block the economic development of the South. As Bradley Parks and Timmons Roberts (2010, p. 146) say at the end of their analysis: 'the fact of the matter is that Southern governments view their interests according to their worldviews and causal beliefs and this appears to be impeding international environmental cooperation'.

Since 2009, the issue of climate justice has continued to be mobilized in different circles. It is worth mentioning the 'Principles of Climate Justice' adopted by the Mary Robinson Foundation (2011), named for its founder, the former President of Ireland and former United Nations High Commissioner for Human Rights. These very general principles are aimed at reintroducing the climate change issue into the framework of human and development rights. From a mixed perspective, between legal and militant, a group of 13 experts in international law and environmental law, led by Pogge and Shelton, presented in March 2015 what they called the *Oslo Principles on Global Climate Change Obligations* (Expert Group on Global Climate Obligations, 2015) (see Box 4.2). The goal of this text was to describe what they thought to be the main moral and legal obligations that states and businesses already had to avoid crossing a critical level of climate change, expressed by the 2°C threshold for temperature rise in the lower atmosphere. In its preamble, it states that:

> Avoiding severe global catastrophe is a moral and legal imperative. To the extent that human activity endangers the biosphere, particularly through the effects of human activity on the global climate, all states and enterprises have an immediate moral and legal duty to prevent the deleterious effects of climate change. While all people, individually and through all the varieties of associations that they form, share the moral duty to avert climate change, the primary legal responsibility rests with states and enterprises.

BOX 4.2 THE OSLO PRINCIPLES: AN OUTLINE

1. The 'precautionary principle' requires that: (a) GHG emissions be reduced to the extent, and at a pace, necessary to protect against the threats of climate change that can still be avoided; and (b) the level of reductions of GHG emissions required to achieve this should be based on any credible and realistic worst-case scenario accepted by a substantial number of eminent climate change experts. The measures required by the precautionary principle should be adopted without regard to the cost, unless that cost is completely disproportionate to the reduction in emissions that will be brought about by applying it.
2. States and enterprises must take measures, based on the precautionary principle, to ensure that the global average surface temperature increase never exceeds pre-industrial temperature by more than 2°C.
3. States and enterprises have the obligation to carry out all the emissions reductions that do not result in additional costs and to refrain from starting activities that involve excessive levels of GHG emissions. The latter include, for example, erecting or expanding coal-fired power plants, without taking countervailing measures, unless the relevant activities can be shown to be indispensable in light of prevailing circumstances, as might be the case in the least-developed countries.
4. Developed and developing countries, as well as enterprises, must take available GHG emissions reduction measures that entail costs if the costs will be offset through future savings or financial gains. Least-developed countries and local enterprises in least-developed countries have the same obligation, to the extent that other entities provide the financial and technical means required without imposing more than a minimal financial burden on the relevant least-developed countries or enterprises.

5. No country or enterprise is relieved of its obligations under these principles even if its contributions to total GHG emissions are small.
6. States and enterprises must comply with the obligations set out in these principles even if relevant national law or international agreements, whether existing or later promulgated, set lower standards and thus would result in less reduction of GHG emissions than required by these principles.
7. The obligations of states are common but differentiated. The least-developed countries have no legal obligation to reduce emissions at their own expense.
8. States must refrain from providing new subsidies, aid, credits, grants, guarantees or insurance for installation of major new facilities or major expansion of existing facilities that will result in the emissions of unnecessarily high or, in the given circumstances, unsustainable quantities of GHG, either within or outside their territories.
9. Neither high cost nor the lack of financial means can, alone, excuse a state's failure to meet its obligations to achieve GHG reductions or constitute a defence against legal sanctions that may be imposed because of such a failure.
10. States must regulate GHG emissions in their jurisdictions or under their control to meet their obligations set forth in these principles.
11. States must accept the jurisdiction of independent courts or tribunals in which the state's compliance with its obligations as set forth in these principles can be challenged and adjudicated. They must participate in these proceedings in good faith and ensure that such proceedings are fair and efficient.

Source: Selective synthesis based on proposals of the Expert Group on Global Climate Obligations (2015).

4.2.3 Experts

The debate between experts, involving ethicists and economists, gave rise to a rather broad spectrum of proposals mainly focusing on the issue of the allocation of emissions rights and the burden associated with mitigation and adaptation. For the 2007 Intergovernmental Panel on Climate Change (IPCC) report (Halsnæs et al., 2007, pp. 144–7), a first dichotomy opposes the utilitarian approach of economists to rights-based approaches. Each one gives birth to various proposals aimed at maximizing global well-being versus meeting basic needs or presumed individual rights to the global atmosphere.

In 1990, for the first IPCC–OECD Workshop on economic and financial measures, in a contribution presented by France, Olivier Godard and Jean-Charles Hourcade (1990) proposed to take account of three variables – recent GHG emissions, per capita gross domestic product (GDP) and adult population of each country – in order to allocate rights to countries based on evolving weights given to each variable and an abatement goal. In the beginning, the biggest weight would be given to recent emissions, whereas as time passed a greater weight would highlight the other two criteria. These variables were to be combined according to the following formula:

$$r\,[a(E_i/E_g) + b(Pop_i/Pop_g) + c(GDP_{icap}/GDP_{gcap})]\,E_g$$

with a, b and c, weights such as $a + b + c = 1$; E_i, emissions of country i; E_g, global anthropogenic emissions; Pop_g, global adult population; Pop_i, adult population of country i; GDP_{icap}, per capita GDP of country i; GDP_{gcap}, global per capita GDP; r, rate of abatement of global emissions by a given target year (for instance, $r = 0.50$ by 2050).

Other multi-criteria proposals have been made by several authors (Ringius et al., 1998; Helm and Simonis, 2001), and a great variety of criteria and formulas have been put on the table, depending on the social value deemed to be predominant. The following are of note (Gupta et al., 2007, p. 770; Okereke, 2010, p. 467):

- Equal per capita allocation (Baer et al., 2000), meaning that each country would be allocated emission entitlements based on its population.
- Equality of sovereign states (Godard, 1992, 2011). If the attribution of GHG emissions rights is comparable to a civic principle, analogous to voting rights, the norm should be to grant equal rights to sovereign states.
- Contraction and convergence (Meyer, 2000), aimed at a process of global decrease of GHG emissions towards an agreed long-term stabilization level for greenhouse gas concentrations ('contraction'), combined with progressive convergence of per capita emission levels of individual countries.
- Basic needs or survival emissions (Pan, 2005; Shue, 2009). Emissions rights would be defined by an assessment of conditions to meet in order to satisfy basic or survival human needs.
- A mix of per capita and per GDP allocation (Gupta and Bhandari, 1999).
- Grandfathering (Rose et al., 1998; Bovens, 2011), which means that reduction targets are based on current emissions or emissions at a given date (1990 in UNFCCC).
- Historical responsibility according to the Brazilian proposal (Rose et al., 1998; Pinguelli Rosa et al., 2003; Rive et al., 2006; den Elzen et al., 2013). Reduction obligations between countries are differentiated

in proportion to countries' relative share of causal responsibility for climate change (contribution to the increase of average global temperature in the lower atmosphere).

- Property rights and market efficiency (Stewart and Wiener, 2003; Posner and Weisbach, 2010). Create a world carbon market to achieve the lowest net cost and maximize global well-being.
- Ability to pay (Caney, 2005b; Moellendorf, 2014). Emission reductions based on ability to pay (levels of welfare or wealth).
- Equal relative mitigation costs (Rose et al., 1998). Allocation of reduction targets to countries designed to equalize relative welfare loss (mitigation costs/GDP ratio).
- Interest-based mutual advantage (Posner and Weisbach, 2010). Allocate rights and duties in such a way that no party would be a net loser regarding a no-agreement baseline.
- Kantian test (Okereke, 2010; Godard, 2016). Countries choose levels of mitigation according to a burden-sharing rule that they would find fair and valid if all countries applied this same rule.
- Sectoral-based national targets with equal efficiency improvements in production and convergence towards equal per capita emissions rights in the domestic sector (the triptych approach) (Gupta et al., 2007).

When discussing the fairness of alternative emission allocations and that of burden-sharing formulas, taking climate impacts and reduction costs into consideration, Benito Müller (2001) concluded that there is no solution that can be considered as the right and fair one far out in the future for every country. Thus, 'what the negotiation architecture does need is an effective and fair procedure to deal with the distributional disputes which may arise in morally ambiguous contexts' (ibid., p. 286).

In fact, since 1992, discussions on how GHG emissions quotas should be allocated between countries have focused on three variables: population, GDP and past emissions, either during a historical period or in a reference year (for instance, 1990). Whatever the ethical norms of justice considered, it is the choice of one among these three variables that is empirically the most important for the respective shares of emission rights between countries (Grasso, 2012, pp. 679ff.). Beyond that, what stands out is the ambiguity attached to those variables, since each of them can be used as a foundation of rights or, alternatively, of obligations and duties.

Take the variable 'population'. In the name of equal individual rights to a common natural resource, emissions quotas could be allocated in proportion to the population of each country; conversely, according to states' responsibility for their population policy, a responsibility that Rawls

(1993a, pp. 47–8) acknowledged, smaller allocations could be assigned to countries with high population growth; this is the practical effect of proposals that determine country allocations based on a past population size, in 1990 for example, which by construction will not take account of subsequent population growth.[3]

Take GDP. It is an indicator of the economic activity of a country and can be seen as the synthetic expression of the physical demands of production, including the right to emit GHGs as a technological requirement of production, at a specific time, given the techniques incorporated into existing productive capital. In many historical cases of natural resource allocation (land, access to water, and so on), the right to produce includes access to the collective resources required to achieve production (Raymond, 2003). From this perspective, focusing on the economic needs of production, emission rights should be allocated to countries in proportion to their respective levels of a GDP-based index, assuming a regular renewal of the procedure to consider differences in the economic dynamics of different countries. Conversely, we may choose to consider that the higher the level of per capita GDP, the higher the countries' capacity to reduce emissions; accordingly, they would be asked for proportionally higher GHG abatement efforts than countries with lower per capita GDP. The emerging and developing countries have adopted this 'capacity' theory to define the obligations they attribute to the developed countries, and the 'needs' theory for themselves. Because they present their economic development as an absolute priority that should not be compromised for any environmental motive, they have long stated that they cannot accept any constraints on their GHG emissions, a position which began to evolve only in the context of the 2015 Paris Agreement.

Finally, past emissions levels are central to the opposition between those who, according to the idea of the 'historical responsibility' of states, postulate a distribution of rights that is inversely proportional to historical cumulative emissions, and those who believe that past emissions, which were produced legally and legitimately, have given rise to a form of customary rights that should be acknowledged. This second interpretation delivers the implicit meaning of the 'grandfathering' allocation method. This term is used in the literature to designate a procedure for the allocation of rights or quotas that is in proportion to past emissions or usages at a given date. This grandfathering approach is parallel to the idea that when individuals and communities have made a place their habitat, they have also acquired an interest in the environmental quality and ecosystem services of that place. One example would be people living by a river and having, for this very reason, the right to maintain this resource for their own usage against those who, for example, wish to divert the course of the river (Piguet, 2014, pp. 418–19). In that vein, some moralists who disapprove of the use of grandfathering for GHG emissions quotas nevertheless support the assertion of the right of local populations to climate stability in order to preserve the benefits they obtain from natural resources and the ecological conditions they deem essential for the exercise of their fundamental rights to life, health and livelihood (Caney, 2009b, p. 166f.; Shue, 2011, p. 295). Here we have a direct conflict between two types of claims regarding acquired rights to the environment.

In the context of climate negotiations, some countries have at one time supported two additional bases for allocating rights. Both are linked to a major dependency that constrains countries' capacity to reduce GHG emissions.

The first base is the size of the territory. The population/land density index and the extent of a state's sovereign territory have significant impacts on land use and transportation needs, and therefore on derived GHG emissions; *ceteris paribus*, small, dense countries should emit significantly less than large ones, independently of states' sovereignty over the use of the air column above their territory (see below).

The second base is the economic dependency on fossil energy sources. When countries generate considerable wealth through the exploitation of fossil energy sources, or when consumer countries have no access to alternatives for meeting their energy requirements, these specific constraints based on economic needs should be taken into consideration, as claimed by representatives of Australia and the Organization of the Petroleum Exporting Countries (OPEC).

The scope of proposals clearly produces very different outcomes (see Table 4.2) in terms of what is considered a fair and just way to tackle the same situation. Some would involve a complete change in world macroeconomic and geopolitical equilibriums. More generally, differences of opinions on climate justice can be traced back to key proposals on concepts and applied norms of justice. There are such conflicting views on key concepts and key variables of the climate nexus that negotiations are fraught with complexity and slowed down by the difficulty of identifying clear common ground of understanding.

Nonetheless, the various competing conceptions for fair allocation are all rooted in argued moral theories. Were these theories simply proposing a guide for personal behaviour, this pluralism would not matter so much, and would simply contribute to social diversity. But given the global public dimension of the issue and the necessary uniqueness of the international set of constitutive and allocating rules,[4] the complexity of the problem remains unaffected: what should be done with this pluralism of conceptions of justice to achieve a single, fair and just international climate regime?

There are three possible solutions. The first would undertake a critical

Table 4.2 Comparison of emission rights allocations with four criteria

Country	Size (% of global land)	Population 2014 (% world pop.)	GDP 2011 (% world GDP)	GHG emissions 2010 (% global emissions)
Russia	11.4	1.99	2.64	3.38
Canada	6.7	0.49	2.46	1.57
United States	6.5	4.42	21.51	11.57
China	6.4	18.93	10.15	22.32
Brazil	5.7	2.83	3.43	3.23
Australia	5.3	0.31	1.92	1.26
European Union	3.0	7.14	25.08	8.05
India	2.2	17.26	2.55	5.37
Argentina	1.9	0.60	0.77	0.63
Kazakhstan	1.8	0.25	0.26	0.63
Algeria	1.6	0.54	0.27	0.34
Dem. Rep. of Congo	1.6	1.08	0.03	2.22
Mexico	1.3	1.68	1.62	1.31
Saudi Arabia	1.3	0.38	0.93	0.99
Indonesia	1.3	3.54	1.17	3.88
Sudan	1.2	0.49	0.09	0.39
Japan	0.25	1.77	8.19	2.46

Notes: The 17 countries included account for 60 per cent of the world's land. While for six countries (China, Brazil, Argentina, Mexico, Saudi Arabia and Indonesia) the maximum difference between the outcome of criteria is less than 1:3, it reached a maximum of 1:74 for the Democratic Republic of the Congo, then 1:33 for Japan, 1:14 for Canada, 1:9 for the European Union, 1:8 for India, 1:6 for Algeria and 1:5 for the United States. The extremes usually oppose size and population; recent emissions and GDP generally offer intermediate values, which could pave the way for a compromise. However, observe that China emits four times more GHGs per unit of GDP than the United States.

Source: Data compiled from (a) List of countries and dependencies by area, Wikipedia (size); (b) UN demographic statistics (population); (c) World Bank Data (GDP); (d) den Elzen et al., 2013 (GHG emissions).

examination of alternative theories in order to dismiss most of them and elect the 'true' one; it would assume that pluralism is only a transitory state due to unfinished reflection.

The second would explore the possibilities for shaping a domain of common understanding, by tracing partial convergence among proposals to obtain majority support for a given set of rules.[5] A variant would be to search for consensual meta principles compatible with a range of doctrines; this approach echoes the aim of Rawls's (1993b) political thinking in *Political Liberalism*: finding proposals for just institutions that could be unanimously approved by subscribers to various social values and conceptions. In the context of climate negotiations, this exercise would generally imply mitigating the requirements of justice and those of political acceptability. Several contributions by economists (Bohm and Larsen, 1994; Young and Wolf, 1992; Ringius et al., 1998; Rose, 1990) have attempted to find a combination of allocation criteria that provides outcomes deemed acceptable to all parties.[6] In such cases, the status of the criteria of distributive justice is reversed: they would cease to provide the legitimate foundations for rules and burden-sharing, but become only instrumental to the overriding criterion of political acceptability.

The third possible solution screens theories in order to test their suitability to situations with the purpose of selecting the most appropriate doctrine and criterion for the specific situation under consideration; the solution is not sought within the pure abstract moral theory, but in the appropriateness of 'universes' that are stylized by theories to the main features and components of the situation in which nation states have to find a cooperation regime for tackling climate change. This is the path I will explore in Chapter 8.

4.2.4 From the Cosmopolitanist Side

Authors such as Shue (1999), Eric Neumayer (2000), Darrel Moellendorf (2002, 2014), Simon Caney (2005a, 2005b, 2009b), Axel Gosseries (2007) and Vanderheiden (2008) are among the leading proponents of a cosmopolitanist approach to the climate justice issue. The starting point of their argument is the assumption of a single human moral community to adopt fair and equitable rules based on a principle of equality of all individuals. Gosseries (2007, p. 280) explains that:

> It is cosmopolitan in the sense that we assume that justice between members of different countries does not morally differ in a significant way from justice among co-citizens or co-residents . . . we are each supposed to have as strong obligations to a foreigner living on another continent as we have to a co-citizen living next door.

In the case of the climate, the initial premise of equality of individuals is associated by some authors with two other assumptions. In the first – which is inherited from seventeenth-century thinking, especially that of John Locke, who considered that the Earth was given to humankind by God – the whole spectrum of natural resources is the common property of humankind. From this premise, these authors often jump to the idea that every human being should enjoy the same rights to these resources, through space and time. I will discuss this idea further.

As for the second assumption, statements and conclusions established for individuals are supposed to be fully transposable to groups and political communities, including states, by simple arithmetic aggregation. At odds with the premises of individualist ethics, the idea of collective responsibility arouses a rather general reluctance from a moral point of view. To overcome this problem, Shue put forward the practical need to make recommendations in a non-ideal world, which is a rather poor justification. The non-cosmopolitanist author Miller (2007) tackled the issue more seriously by elucidating the diverse conditions under which individuals take part in the political processes in the society – he considers a nation – in which they live, and concludes that:

> Where nations act in ways that impose burdens on themselves or on others, responsibility for such burdens falls on every member, even on those who opposed the decisions or policies in question. The argument turns on the sharing of beliefs and attitudes that characterizes national communities, and on the benefits that membership brings with it. (ibid., p. 133)

We see that what enables Miller to give legitimate meaning to the concept of collective responsibility are specific features originating in the national links between citizens, something that cosmopolitanists put aside as contingent and arbitrary. This abstract individualism leads them to a contradiction between the foundations they give to their proposals, and the institutional solutions they eventually mobilize, which all require agreement and action by governments of nation states and international organizations, in other words collective bodies.

Based on these premises, two intellectual operations were initiated by proponents of the 'historical responsibility' of countries: retrospective determination of the fair share of each country, and determining countries' future rights according to their population.

Retrospective determination of the fair share of each country

The first step is to reconstruct the history of the emissions of each country in the world – in practice they have focused on emissions of the major developed countries – since the beginning of the industrial era (1750 or 1850, depending on the author, without any significant impact on cumulative data). The second step is to determine an overall budget of cumulative emissions consistent with the goals of climate protection (for instance, the 2°C target). The third step is to compare the actual cumulative emissions of each country with what would have been the 'just' egalitarian distribution of CO_2 emissions rights to countries based on their population since 1850. This comparison determines the 'excess' emissions from the developed countries, which is considered to be the foundational source of historical injustice towards the developing countries. Having taken more than their 'fair share' of the limited carbon budget, the developed countries have reduced the remaining share available to the South, and violated the South's rights to use the atmosphere. The current populations of the Northern countries, as the heirs and beneficiaries of this historical injustice, would have the duty to discharge the debt jointly accumulated by their ancestors and themselves.

These issues of 'historical responsibility' and the climate debt of the developed countries have been the permanent background to many proposals by NGOs and governments in the South, but also in Northern militant circles, such as those involved in the Environmental Justice Organizations, Liabilities and Trade (EJOLT) programme funded by the European Union (Warlenius et al., 2015a).

Despite its evidence at first sight, this argument has not been unanimously accepted, either by Southern experts or by the proponents of cosmopolitanism. Thus, a prominent figure on the Brazilian scientific and political scene, the physicist José Goldemberg, who was the organizer of the Rio Earth Summit in 1992, has clearly distanced himself from this rhetoric (Goldemberg and Guardabassi, 2012, 2015). Firstly, from a quantitative and empirical point of view, CO_2 emissions prior to 1990 will soon count for only a small proportion – one-third – of what will be the total cumulative emissions from 1850 to 2020, and the focus on the 'historical responsibility' of the North is therefore an unwelcome distraction from what really matters. Secondly, the issue of 'historical responsibility' has mostly been a pretext used by the emerging and developing countries to justify their inaction, whereas these countries can now use modern energy technologies, particularly for renewables, that were not available to the countries that embarked on industrialization in the nineteenth century.

Taking the same path of rebuttal, the philosophers Miller (2009), Risse (2012) and Moellendorf (2014) challenged point by point the possible moral foundations of the idea of the 'historical responsibility' and 'climate debt' of the North (fault liability, strict liability, liability of beneficiaries of an injustice). In the same way, Frédéric-Paul Piguet (2014, p. 415), noted for his radical proposal on the recognition of a general principle of 'forbidding harm by negligence' and the attribution of a 'hierarchically superior common good' status to the preservation of the integrity of the biosphere, explains that: 'Morally, it is difficult to blame a public authority for emissions prior to 1985, since no consensus existed at that time . . . It does not seem wise to require compensation for damages for past emissions prior to 1990'. I will discuss these issues in Chapter 6.

Determining countries' future rights according to their population

Most representatives of the emerging and developing countries have argued for the future carbon budget to be allocated in proportion to the population of each country. This idea also underlies the 'contraction and convergence' proposal, which is aimed at organizing a transition process towards equality over several decades.

This procedure linking emissions rights to country population size has been endorsed by many moralists who have written about the climate threat, including Peter Singer (2008, p. 682), Shue, Caney, Gardiner and Piguet, as well as by economists such as Neumayer (2000). This position is directly related to the individualistic postulate of the cosmopolitanist approach, which is opposed on this point to the communitarian or utilitarian approaches, but also to the Rawlsian one. The latter, for instance, certainly considers individual liberties as the first value, but does not adopt the concept of 'natural rights' to natural resources as foundational; Rawls's reasoning on equality of opportunity and the difference principle concerns statistical groups rather than individuals.

The direct effect of an individualistic ethic on the interpretation of a population-based conception of an equitable distribution of emissions rights was demonstrated by Snorre Kverndokk (1995): for such individualist ethics, the equitable sharing of emissions rights can only be envisaged using morally relevant parameters for individuals. This is not the case of collective variables such as GDP or country size. But this individualist approach encounters difficulties when we consider that most envisaged procedures for the allocation of emissions rights are aimed at distribution to governments, with a view to developing exchanges of quotas with other countries. How can they justify such procedures in the name of the basic rights of individuals? The problem remains unresolved by cosmopolitanist individualism.

The Indian researcher and activist Anil Agarwal pioneered this issue of equal per capita allocation in a text denouncing the 'neo-colonialism' that transpired, in his eyes, from the first calculation of net GHG emissions per country made in 1990 by the World Resources Institute (Agarwal and Narain, 1991). To account for the sequestration of atmospheric carbon by the oceans and terrestrial ecosystems and to determine net emissions,[7] calculations by the Institute attributed these 'carbon sinks' to countries in proportion to their level of gross emissions. Agarwal asked for the benefits of GHG absorption by sinks not situated in state territory to be allocated to all countries in proportion to their population. This was the beginning of a well-publicized controversy, and a landmark for all subsequent proposals on the individual equality of GHG emission rights.

In academic circles, not all cosmopolitanists defend this principle of equal emission rights per capita. Caney (2011) undertook a systematic critique of this principle, while reaffirming its commitment to an egalitarian approach focusing on meeting the basic needs of people. Starting from the premise of the common ownership of the Earth, Risse (2012, pp. 194–8) nevertheless contested Singer's (2008) assertion that the principle of an equal share to every human being can be deduced from this premise: some regulations are needed to organize usage of this common ownership, but not necessarily in the form of equal sharing. Vanderheiden (2008) proposed the equity criterion to regulate the allocation of emissions reduction efforts, and that of liability to govern compensation for damage and obstacles due to past emissions. To define equity, this author adopted the Rawlsian 'Principle of Difference', but expanded internationally, as advocated by Moellendorf (2002) and Pogge (2004). Consequently, wrote Vanderheiden, we should account for differences in the level of survival emissions by country and for the investments made by some countries in new carbon sinks that would increase the ability of natural systems to absorb GHG emissions.

More recently, without relying on his earlier analyses for this point, Moellendorf (2014) suggested focusing the distribution of GHG emission rights on the capacity of each state to promote emissions reduction activities. His view was based on the simple idea that the higher a state's GDP, the higher its capacity for emissions abatement. With another cosmopolitanist proposal that diverges from the equal per capita principle, Shue (2009, 2014) called for the adoption of a principle of lexicographical priority for so-called 'subsistence' emissions rights. His estimation was that the subsistence level was around the current average level of CO_2 emissions for India (1.6 tCO_2 per capita in 2013). Others put the threshold at 1.2 tCO_{2eq} per year (den Elzen et al., 2013). Considering the future growth of the world population – the population of 2000 could double by the twenty-second century – and the future economic development expected for the twenty-first century, this principle of subsistence would, in Shue's opinion, almost completely exhaust the carbon budget available.

4.2.5 From an International Viewpoint

For advocates of justice between nation states, the existence of these states takes on a central significance, not only factually, but also in terms of political and moral values. Humankind is not composed of a single polity with 7 billion members, but is arranged into a society of almost 200 sovereign states. Any proposal for a new international rule must pass the test of the free consent of each state. Furthermore, a key condition for international climate justice is the equitable access of all states to international negotiations and governance of the common institutions, such as carbon funds or various committees. Consequently, for a state to want to play the card of cooperation with others, it must be confident of obtaining clear advantages in terms of its projects and values. The regime to be set up cannot be a punitive one, in that it would represent a punishment for certain states on grounds of their past behaviour prior to the agreement (new international standards cannot be retroactive). Moreover, the benefits of cooperation must be fairly distributed; at least, they must not be captured by one or another of the parties.

Finally, that which belongs to the core of sovereignty is not allowed to be threatened by rules of international justice, even as a premise for arguments such as that underlying the thesis of 'climate debt'. A political entity such as a state is by nature regarded as an independent entity that is able to be self-sustaining in the long run. Exchanges and interactions with other states should not be such that a foreign power can hold the state in its hands, in which case it would no longer exist as such. This, as Chauvier (1999, p. 15) said, is what ensures that states are not, in their mutual relationships, in a situation of social cooperation like that of individuals within a given political society. Individuals are in a situation of widespread and multidimensional interdependence with their fellows; and in a situation of dependency on the state authority, even if in a democracy, citizens contribute to the formation of the general will. It is different for states. Control of a territory is not just one good among others that states would value but could be deprived of without losing their identity: it is an essential part of their autonomous existence; a state is a territorial entity (Chauvier, 1999, p. 37).

From the perspective of sovereignty, territory includes the ground, below the ground, and the airspace that rises vertically from within national borders, here expanded to include territorial waters. Sovereignty over this air column extends up to the limit of the atmosphere and stops at the beginning of outer space governed by the Space Treaty of 1966. This represents an altitude of around 100 km above sea level (Verhoeven, 2000, pp. 583–4). It would be inconceivable, being contradictory to the very idea of political sovereignty, that any use of the atmosphere above the territory of a country could be under the control of a foreign state or an entity claiming to represent humankind. It is not by chance that international law recognizes the principle of the permanent sovereignty of states over their natural resources. Although the legal status of the air is rather unclear – it is neither a 'shared natural resource', nor a '*res communis*', nor the 'common heritage of humanity' – policies aimed at protecting its quality were recognized in 1996 as the 'conservation of exhaustible natural resources' by the World Trade Organization's Appellate Body under the Dispute Settlement Body. Justice between sovereign states rests primarily on mutual recognition of the attributes of sovereignty, including possession of the natural resources of a territory, and usage of 'territorial' air.

Consequently, the international approach to justice provides no outlet for a requirement concerning the international redistribution of natural resources placed under the control of states. However, several authors acknowledge the existence of international obligations that can modulate or supplement a primary allocation of GHG emissions rights that would have been based on the territorial size of states. Three reasons are put forward (Rawls, 1993b; Miller, 2007):

1. providing for the basic needs of people whose survival or subsistence is threatened;
2. assisting fragile states in their struggle to achieve the economic and social conditions to establish or restore fair and efficient public institutions capable of ensuring their self-determination;
3. repairing internationally acknowledged historical injustices.

Thus, there is a legitimate basis for policies of financial aid, technology transfer and capacity-building for the benefit of the least-developed countries or those most vulnerable to climate change.

Some of the demands of international justice between sovereign states were reflected in the UNFCCC text. Acknowledging the right of all countries to sustainable development, it reaffirmed 'the principle of sovereignty of states in international cooperation to address climate change', and that 'states have the sovereign right to exploit their own resources pursuant to their own environmental and developmental policies, and the responsibility to ensure that activities within their jurisdiction or control do not cause damage to the environment of other states or of areas beyond the limits of national jurisdiction'. No signatory to this convention can claim a right to a development style that would be insensitive to its environmental costs to others. But none can claim to have rights to the natural resources of others, including 'territorial' air.

Some of the flagship proposals made in the context of climate negotiations are clearly not consistent with that foundation. This is the case for the premise of making the atmosphere the common property of humankind to which all human beings should have the same rights, whether they belong to past, present or future generations. Before the international society of states agreed to new rules in 1992 under the UNFCCC, the different uses states made of their airspace were legitimate. The idea that some countries have made excessive historical use of the atmosphere, regarding a norm of equal per capita distribution supposed to have been valid since the beginnings of history, is pure nonsense here. However, the issue of possible damage caused to the environment of other countries by GHG emissions from a given country remains valid, since the GHGs emitted are not contained within the territorial space in which they are released, but spread into the air flowing around the planet.

This rule of 'equal per capita GHG emissions rights' would also contradict the principle of sovereignty in another respect: it would lead to huge disparities in terms of the economic costs to be borne by the states and impose huge net costs on some of them, contrary to the requirement of mutually beneficial cooperation. The full implementation of these proposals would spell disaster for most of the industrial facilities and transport systems of countries such as the USA (see Box 4.3). The economic and political weakening that would ensue is so significant that it would

BOX 4.3 THE US CASE

The population of the USA in 2014 (318 million) amounted to 4.42 per cent of the world population (7.2 billion). The country's cumulative emissions over the 1850–2010 period accounted for 18.9 per cent of the global total, that is, 482 $GtCO_{2eq}$ out of a total of 2550 $GtCO_{2eq}$. But total cumulative global emissions should have been capped at a lower level of at least half that amount to avoid the climate threat, which would have entailed a 1275 $GtCO_{2eq}$ cap. With the equal per capita criterion applied to the 2014 population, the GHG budget of the USA over the 1850–2010 period, assuming the same population rate, would have been limited to 56.3 $GtCO_{2eq}$. This country should have avoided more than 88 per cent of its emissions for more than a century and a half (a division by 9) to remain within its 'rights' to historical emissions thus calculated. In absolute terms, its historical reparation obligation would amount to 426 $GtCO_{2eq}$.

Turning to the future and adopting a target of halving global GHG emissions over the 2015–25 period, the allowable annual global budget would be about 25 $GtCO_{2eq}$. Of this budget, the US would obtain a right proportional to its population, which would give this country an annual emissions cap of 1.1 $GtCO_{2eq}$, assuming a constant relative demographic weight. In reality, current emissions from this country amount to 6.5 Gt. In addition to its obligation to make reparation for its past carbon intensity, this country should divide its future emissions by six.

The combination of these two requirements, called for by several proposals including those of Vanderheiden (2008), would imply the following figures, in order of magnitude, for a social cost of an avoided tonne of CO_2 emission estimated at between US$10 and US$60: between US$4260 billion and US$25 560 billion for the past period prior to 2010, in other words between 25 per cent and 152 per cent of US GDP in 2013, and between US$54 billion and US$324 billion annually in respect of future obligations. In fact, the economic impact would be higher than is suggested by these figures. Faced with the almost total loss of its ability to emit CO_2, the country would suffer almost total paralysis of its energy system (coal, oil and gas extraction, petrol for vehicles, and coal and gas for power generation) and its economic activity in many areas. The purchasing power of its consumers would fall dramatically, before costly investments in alternative techniques could re-establish an operational energy system.

undermine their sovereignty, including from the point of view of military capabilities to defend their territory.

If we reject the criterion of equal per capita allocation for the international distribution of emissions quotas, but do not go as far as proposing the size of territories as a distribution base, which suggestion would be in line with an international justice between states? Miller (2009, p. 145) suggested renewing with the classical maxim 'from each according to his abilities, to each according to his needs'. This should involve two phases. First, identifying societies where poverty is endemic and cannot be overcome by internal redistribution between the richest and the poorest of these societies. No GHG emissions constraint would be imposed on them if they use the most effective techniques to reduce their impact on the climate. For other states, the objective should be the equalization of sacrifices in terms of GDP percentage points lost due to climate policy. This criterion is inspired by an idea of just burden-sharing from international cooperation, without any special redistributive purpose, except for exempted least-developed countries.

However, one limitation of this proposal is that it does not consider differences in the benefits obtained by countries from vigorous action against climate change (countries with access to the sea will benefit more from the limitation of sea level rise than landlocked countries). Moreover, while challenging the foundations of 'historical responsibility', Miller admitted the involvement of liability for damages and adaptation costs for vulnerable populations in respect of emissions produced since 1990. Once estimated, these costs would be allocated to countries in proportion to their recent GHG emissions. Miller warned that the international regime thus defined should interfere as little as possible with domestic policies, particularly with respect to the distribution of efforts and constraints between social groups. Such matters fall within the sovereignty of each country as a political community.

Having severely criticized the norm of equalization for individual emissions rights, Posner and Weisbach (2010) argue in a book that is often criticized by moralists,[8] but is largely welcomed by economists, for a solution ('International Paretianism') that would guarantee that no state would be a net loser in a new international regime in terms of economic well-being.

The review of proposals made on behalf of global climate justice highlights the diversity both of starting points for theories, arguments and proposals, and of final recommendations. These are not all equally convincing. Two distinct problems have been identified which, although partially connected, solicit the idea of justice:

1. How can GHG emissions rights be distributed equitably among countries within a future carbon budget that will be severely restricted? Are both ideas – of distribution proportional to the population size, and retroactive budgets shared according to an egalitarian norm over the historical period since 1850 – in any way defensible?
2. Should the current and future state of the climate – which is presumed to be associated with damage experienced by the population, additional costs for adapting to the new climate and additional barriers to future economic development – require the payment of compensation to victims, and if so, who should pay?

When considering these two questions (the allocation of rights, compensation for damage), two subsidiary questions arise:

3. Is it relevant to consider as such, in the name of international distributive justice, huge economic inequalities between rich and poor on this planet? Should these be given a place in a new international climate protection regime, or be considered separately as a developmental issue?
4. How should the relationship between present generations and distant future generations be described and considered?

The following chapters will provide the opportunity to examine these questions.

NOTES

1. This statement is true for CO_2, but not for all GHGs or their precursors. Short-lived gases also have specific impacts on regional climates.
2. Note that the French population of consumers increased over the period in question by nearly 10 per cent, with 5.6 million more inhabitants.
3. To take account of the potential feedback of an allocation rule on population dynamics and policies, Michael Grubb (1995) mentioned the possibility that CO_2 quotas could be allocated by referring to the adult population rather than to the total population.
4. International rules may be designed in such a way that they preserve sovereignty over the choice of purely domestic regulatory regimes, embracing both the choice of policy instruments (taxes, emissions trading, planning, information, and so on) and the distributive dimension. But when autonomy is preserved for domestic rules, it raises limitations for the justification of certain concepts and bases proposed for international burden-sharing. This is the case, for instance, of the per capita allocation rationale, which mobilizes arguments that would entail equal per capita allocation on an individual scale.
5. This is what Shue (1999) suggests, but without considering political acceptability in international negotiations. Sen (2009) advocated this approach to seeking points of convergence between several moral and political constructs in order to make progress in

correcting huge injustices, instead of exacerbating differences and the incompatibility of premises, and searching for the perfect solution.

6. This approach was defended by H. Peyton Young and Amanda Wolf (1992) as follows: 'Equity is a complicated, nuanced business that cannot be reduced to simple recipes like grandfathering present emissions or handing out equal numbers of permits to everyone. Equity involves balancing various considerations – past responsibility, present practice, ability to pay, and need . . . equity is not some utopian ideal dreamed up by philosophers; it is the tool that practical negotiators use to forge consensus'.
7. Box 4.1 recalls that about 45 per cent of CO_2 emissions remain in the atmosphere after a few decades, with the rest being absorbed by oceans and terrestrial sinks.
8. One exception is Nussbaum (2013), who writes: '*Climate Change Justice* is an impressive and sensible book . . . it shows real care about philosophical theories of justice and takes their claim seriously'.

5. The puzzle of intergenerational equity

Due to its specific time dimension, the threat of global climate change necessitates considering and articulating different time horizons, from the very long term (several millennia) to the short and medium terms, which are relevant for mitigation and adaptation policies. It is not only that some events of concern may occur in the very distant future while others are happening in the present time; the challenges concern more puzzling issues. Firstly, we must tackle pervasive uncertainty about causal chains and actual expressions of climate change, but also about the state of societies, with their social values, lifestyles, technologies and social structures. Secondly, we, the present generations who must make policy decisions, face the issue of how to consider the unborn. 'Future generations' are those potential people in whose name so many moral discourses are elaborated. We cannot speak to them to learn about their preferences. We do not know in what kind of world they will live. We are not even certain that they will actually exist. In this context, how and to what degree should current generations be concerned with the long-term effects of the footprint of economic growth on the planet's health? How should we understand the requirements of intergenerational equity?

In this chapter, I first examine the debate on discounting to show the true and the false. Next, the main issues of intergenerational relationships are considered in order to test different moral propositions put forward in the name of intergenerational justice, with a special focus on the 'non-existence' and 'non-identity' problems. I also question the foundations of duties to the dead and the unborn, which may threaten the freedoms of living generations. I end by considering potential ways to address unresolved problems.

5.1 THE UTILITARIAN ANSWER

5.1.1 The Rationale of Discounting Future Values

A common ground

Belonging to utilitarianism, the standard economic response to these questions, using the tools of applied welfare economics – in other words, cost–benefit analysis – is to assess and evaluate possible climate-related damage throughout long-term economic growth trajectories. To this effect, the range of goods of human interest and their relative prices should be determined by internalizing external diseconomies and correcting other market imperfections. Time-indexed values should then be discounted to be able to compare values at different times.[1] Within this utilitarian framework, discounting relies on the price vector giving the present value of the numeraire through time.[2] The discounted sum of climate change-induced damage estimates associated with a given growth trajectory and a climate and energy policy is then compared to that of the costs of various actions or policies aimed at avoiding or minimizing the damage associated with this trajectory to different degrees.

The whole discounting procedure requires the translation of non-market effects into market equivalents using monetary assessments, of uncertain impacts into certainty-equivalent ones, and of future values into present equivalents. The distinction between these three steps should be highlighted, since some critical reviews of discounting are rather confusing, for instance by underlining difficulties in relation to damage assessment as a sufficient reason for rebutting discounting as such, or by calling for the use of different discounting rates for different types of goods.

There is widespread agreement among welfare economists that a discussion of discounting rates should arise from Frank Ramsey's well-known simplified equation in an optimal economy: $r = \rho + \eta g$, with r, the marginal productivity of capital; ρ, the rate of pure time preference, or the utility discount rate, or the rate of impatience; η, the absolute value of the elasticity of the marginal social value of consumption; and g, the per capita annual rate of consumption growth. The expression $\rho + \eta g$ can be designated as τ, the social rate of preference for present consumption (SRPPC). It reflects the appreciation that a representative agent – often called a social planner – is supposed to make, in utility terms, of different proposed combinations of renunciations and gains in consumption distributed over time. The goal of society is supposed to be maximizing total utility extracted from consumption, an expression that includes everything that can be used or consumed in the world as instrumental responses to human needs and desires.

Ramsey's equation relates the dynamics of production, reflected directly in r and indirectly in g, to those of preferences for consumption and savings, the latter being understood as postponed consumption, reflected in ρ and η. It calls for moral positions to consider economic conditions of choices and their consequences.

Understanding the pure time preference

The usual presentation of τ justifies the pure time preference ρ through psychological observations: when given a choice, individuals prefer an immediate pleasure to a delayed one when these pleasures have the same 'value' and are altogether certain. For a marginal displacement of consumption plans for one individual, this may be understandable; it is not unreasonable to prefer to enjoy oneself by satisfying a desire tomorrow rather than in 40 years' time. Observations of household behaviours could provide a value of between 1 and 2 per cent for ρ, whereas climate models of William Nordhaus alternate between a value of 3 per cent in 2006 and 1.5 per cent several months later (Nordhaus, 2007). However, from the viewpoint of society, the argument must be changed: issues of intergenerational equity do matter, since the time horizon of society is not the same as that of individual people, even if, as Thompson (2009, Chapter 4) says, they have 'lifetime-transcending interests'; there is no justification for giving priority to the present utility of consumption over the future utility, and a fair equilibrium should be found over time.

At this point, it is important to address a specific difficulty of utilitarianism: being a sacrificial doctrine (Rawls, 1971 [1999]). In postulating an infinite time and an infinite number of generations, Tjalling Koopmans (1960) demonstrated that ρ had to be strictly positive. Otherwise, under the welfare-maximizing criterion, the almost total sacrifice of present generations' consumption would be required in order to benefit an infinite number of future generations. The variable ρ therefore has the function of limiting the effect of the sacrificial rationale in an intellectual setting – utilitarianism practised by economists – that gives no place to concepts of basic rights or freedoms, to which Rawls, in contrast, attributes a superior lexicographical value. In the Nicholas Stern Review on *The Economics of Climate Change* (Stern, 2007), ρ is maintained positive at a value of 0.1 per cent to reflect the probability that future generations do not exist at all.

The issue of inequality between generations

For public choices, the parameter η is particularly ambiguous, since it alone represents three different notions: the degree of social aversion to inequality within a population of contemporary agents, the degree of social aversion to inequality between successive generations, and the degree of social aversion to risk. This confusion originates in John Harsanyi's reinterpretation (1953, 1955) of the Rawlsian veil of ignorance as a situation in which rational individuals make risk calculations based on the possible outcomes of their choice for themselves under a criterion of maximizing their expected utility. Thus, attitudes towards risk and inequality are treated by the mainstream literature as if they were the same concepts: aversion to social inequality is supposed to be the outcome of a person's aversion to the risk of being poor, in other words an egocentric approach to inequality. This assimilation is untenable. Based on empirical studies on public decision-making, Simon Dietz et al. (2008) suggest that in real choices, the aversion to inequality in modern contemporary societies is rather less than 1. However, as a coefficient of aversion to risk, η is generally given values much greater than 1.

If η is set at 1, this means that whatever the wealth of a generation, the utility of a change (increase or decrease) by a given percentage of consumption has the same value for all. If an early generation has access to a consumption of 1000 and a later one to a consumption of 10000, taking 10, that is, 1 per cent, from the first generation would have the same social value as taking 100, that is, 1 per cent, from the second. This can be presented in another way: it would be socially valid to reduce the second generation's access by 100 if this makes it possible to increase the first generation's access by at least 10. Meanwhile, Partha Dasgupta (2008) considers a value of between 2 and 3 as normatively reasonable for η. This implies a considerable social willingness for redistribution from the rich to the poor, but also, in the context of standard growth assumptions, a considerable preference for the welfare of the poorest generations – that is, the present ones – over that of the much richer future generations (in the model assuming g to be positive). With $\eta = 2$, reducing the later generation's consumption by 50 per cent (5000 remaining instead of 10000) would have the same social value as reducing the early generation's consumption by 1 per cent (990 remaining instead of 1000). Such values would create solid obstacles for any policy calling for the least well-off to lose a little more, but do not seem plausible if matched with real redistribution practices in contemporaneous societies, as previously stated, or in the context of international climate negotiations.

Thus, the general idea that the whole construct of discounting is morally unacceptable because it does not consider the interests of the least well-off is ill-founded. If there are doubts about factual assumptions regarding future economic growth and it is thought that the global economic future will involve a decrease in per capita wealth, g will be negative and the value of τ will depend on the relative value of the two components ρ and ηg. With $\rho = 0.1$ per cent, $g = -1$ per cent and $\eta = 2$, for instance, $\tau = -1.9$ per cent, which would mean that present generations should be treated as the richest of all generations to come. With the prospect of economic decline, discounting would continuously increase the weights given to impacts on future generations' utility.

Regarding inequality between generations, it is important to underline and debunk common misunderstandings. Moralists affirm that some items such as loss of human life or access to basic goods, which are required to ensure the respect of fundamental rights, should not be discounted, either because they are not cumulable, or because they represent absolute minima that must not be brought into the balance against other advantageous conditions. Thus, it is said that since all human beings are equal in dignity whatever their birth date, values referring to their life, health, safety and freedoms, and so on, cannot be discounted. We can accept the premise, but not the conclusion when we look in detail at economic methods. For instance, in the case of loss of statistical human lives, if we value – as we should within the utilitarian framework – the losses for the economy as the discounted utility sum of all net incomes that dead people would have generated had they remained alive, such values reflect the level of wealth attained by the generation to which these people belong. If this generation is ten times richer on average than present generations, loss of lives from this richer generation would be valued ten times more than the premature death of people belonging to present generations if discounting is forbidden. It is by discounting not lives as such, but the assessment of costs and damages linked to them, that it is possible to put an equal value to the death of human beings throughout time, whatever the date of death.

A similar type of reasoning should apply to the means of satisfying the basic needs and respecting the fundamental rights of people in each generation. These reflect the future conditions of production and relative scarcity of goods at the date considered. Their present value is defined by the amount that should be invested today in order to produce the economic resources required to provide the necessary goods at stake in the future. The future value of these goods is to be discounted at a rate that reflects the rate of return on investment over the period separating the present time from the future date considered.

Furthermore, the more uncertain that economic growth appears, the more this translates for risk-averse agents into a lesser value of the certainty equivalent of expected future consumption, which results in a lower growth rate, and the more the discount rate must fall (Gollier, 2002), leading to variable discount rates declining in the long term.

Do not forget the structural change in price system through time

To implement discounting correctly, we must find plausible methods to assess the uncertain future costs and benefits of climate change. We must then choose a correct discount rate based on correct values for the three main variables, g, ρ and η. The two tasks are related. To obtain a valid estimate of the value for each good concerned, it is essential to use appropriate time-indexed price vectors for all relevant goods. Thus, we should consider the dynamics of the structural change of a complete pricing system for the whole range of goods provided by a society to its members. This is even more important if the time horizon is in the distant future. Indeed, the relative scarcity of goods will change significantly over the coming decades and centuries. As Thomas Sterner and U. Martin Persson (2008) magnificently demonstrated in the case of climate change, we cannot discuss the discount rate in isolation, but should simultaneously consider the increasing scarcity of natural resources and the growing weight of environmental constraints. Clearly the issue of discounting, from both moral and economic viewpoints, is therefore not so easy to ignore or dismiss. But . . .

5.1.2 An Unreliable Tool

Even if a low discount rate is more appropriate for giving an acceptable, grounded representation of long-term future stakes, the discounting approach remains affected by several sources of practical weakness: (1) the extreme sensitivity of policy choices to a variable for which there is no robust and objective way of coming to a shared assessment; (2) the persistent contradiction of the various aversions assumed to determine the parameter η; (3) the huge uncertainty surrounding future economic growth; and (4) the weak foundations of the assumption of decreasing marginal utility of consumption when humankind is taken as a single intergenerational community: the psychology of satiation invoked for individuals is deprived of meaning at the level of society and humankind.

The conclusion is inescapable: discounting cannot be the central means by which the future of global commons can be decided. It is a tool enabling project assessment within the framework of a sustainable development policy that has already been chosen, based on diverse sources of scientific and technological information on climate issues and the possibilities of achieving a transition to a low-carbon economy.

5.2 DUTIES TO THE DEAD AND THE UNBORN IN THE LIGHT OF FREEDOMS

Intergenerational justice is generally intended to give present generations various duties and obligations. This raises the issue of how we should value the freedom of choice of the living and next generations in an intergenerational context (Sen, 1999). To explore this issue, I will mainly consider the important theoretical reflection put forward by Thompson (2009), who charges each generation with a burden of duties and obligations generated by the projects and desires of past and dead generations, as well as duties to future unborn generations. She does so in the name of the 'lifetime-transcending interests' of human beings, which would give them post-mortem rights over future generations and pre-conception rights of the unborn on living generations.

5.2.1 Lifetime-Transcending Interests as a Source of Rights and Duties

Regarding relationships with dead generations, Thompson justifies the duties of living generations, not only in memorial terms or for issues of private heritage, but by considering how living people should aim to fulfil the projects and desires of the previous generations. Rights of past generations over living and future generations do exist in this respect, says Thompson, since nearly all human beings have not only lifetime basic interests and rights (the rights to life, health, safety, property, and so on), but also lifetime-transcending interests: people cherish certain values, institutions and landscapes, and devote their lives to projects that cannot bear fruit entirely during their lifetime (planting oaks, investing in the education of young children to contribute to the future economic development of a devastated country, and so on). They hope these projects will survive their own death, and this is a key component of the meaning of life for them. These are essential interests and, for this reason, they generate rights that can be opposed to others, and specifically to the next generations.

This same foundation also provides grounds for obligations and duties to the unborn. Since the meaning of life depends on what in the future will become of all the values and goods to which people are attached, people need to: (1) have successors; and (2) make arrangements to ensure these successors continue the endeavours and values they want to see sustained. Thus, in order to have successors, they must preserve specific conditions under which these successors can live and enjoy themselves, and be willing and able to promote the projects, desires and values of their predecessors to their own successors. It is for the sake of the lifetime-transcending interests of living people that the latter have duties and obligations to the unborn, in order to obtain from them respect and willingness to pursue their dreams and actions.

This vision would threaten the fundamental freedoms of present and future generations, should the latter be enslaved to the desires and projects of past generations. Limits are necessary to ensure that all can enjoy themselves and not be reduced to an instrument serving the interests of others. Thompson's theory would have been exposed to harsh criticism, had she not given indications regarding the extent and limit of these specific duties to the dead and the unborn. This is a complex matter. In Thompson's words:

> Individuals do not have a duty to carry on the projects of their predecessors. The mere fact that past people valued something and made an effort to maintain it for their successors is not a sufficient reason for us to value it . . . But we do have a duty to remember and to make an effort to appreciate what our predecessors did for our sake. This means that if we want to reject our inheritance and abandon our predecessors' project, the onus of justification falls on us . . . we might nevertheless have a duty to ensure that our successors have an opportunity to inherit it. (Thompson, 2009, p. 96)

She continues:

> The citizens of each generation are entitled to demand that their successors . . . make their ideas about value – and in some cases, the things they valued – available to succeeding generations so that these more distant successors have an opportunity to understand them and maintain their link with past generations. (ibid., p. 97)

This construct is presented as passing Rawls's test of deliberation regarding just institutions. People meeting under the veil of ignorance to define the set of rules for a just society do not know to which generation they belong or what their interests are. But, as Thomson says, they know they will benefit from developing and pursuing lifetime-transcending interests and, on this basis, have desires that they want their successors to fulfil. Accordingly, it is in their interest to ensure their successors are in a position to fulfil these desires. Being ignorant of the generation to which they belong, they are predisposed to accept a duty of justice 'that requires each generation to provide their successors with the institutions and conditions that will enable them to develop and pursue lifetime-transcending interests' (ibid., p. 51).

5.2.2 Drawbacks

Thompson recognizes that this Rawlsian contractual approach remains difficult to defend, being dependent on the belief that successors will behave as expected. If one generation does not trust that future generations will agree to promote the same values, it will presumably consider itself freed from all obligations to those future generations, and the intergenerational chain of duties will be broken. This is why the ultimate foundation of the system of intergenerational duties and entitlements should not be political, as for Rawls's theory, but moral, with an obligation of balance between demands and duties. 'If we can bind our successors to fulfil some of our lifetime-transcending demands, then we ought to accept an obligation to fulfil legitimate demands of our predecessors' (ibid., p. 53). But what are the legitimate demands of predecessors? Can dead people have demands?

Not all post-mortem demands and desires must be fulfilled by successors. Thompson excludes immoral or illegal demands, but does not explicitly consider cases in which the weight of predecessors on successors may be too heavy and too restrictive for the essential freedoms of the latter. Yet, in Rawls's framework, deliberating people may perhaps wish for their desires and projects to be carried on by the next generations, but they undoubtedly want above all to benefit from essential freedoms to define their own life projects and design their own collective rules. They will not be willing to be constrained by obligations to fulfil all the projects, ideals and values of numerous previous generations. If Thompson is right in saying that successors must make an effort to understand the value systems of their predecessors that explain why they behaved in a certain way, it is not because they have a specific obligation to their predecessors, as in traditional societies; it is because this may enable successors to better understand the anthropology of humankind, to discover valuable goods and ideals that will be beneficial to them, to learn from historical wisdom and to benefit from some components of inheritance.

Even if some generations perceive their own lifetime-transcending interests as imperative laws that their successors should obey, this is no legitimate reason why successors should feel obliged to conform to these desires. What is dubious is the deduction of intergenerational rights from the presumed interests of living generations. The intensity of desires and projects of each generation is no reason for interfering with the respective freedoms of living generations and their successors. The key idea to retain from Thompson's development in this respect is that of an indirect conversation between generations. Such a conversation would undoubtedly require a dual effort of understanding by all parties. Living generations should be concerned by the evolution of the world in which future generations will live, and make efforts of imagination and foresight to gain a better understanding of what is happening in the present time. The same living generations should also be concerned by the conditions in which their predecessors lived, made projects and defended values, in order to gain a better understanding of what all this meant and to what extent this could still be a cultural and social asset for the present time. That being said, this entails no transgenerational obligations of loyalty to the past and to the future as such, since living people are not instruments for others.

Thompson thought that the case of obligations to the past and for the past would be better served by a moral construct than by a contractarian political one, which would be exposed to the threat of suspicion and defection between generations. However, the moral argument based on the balance of duties and demands does not provide a more solid foundation: without impairing the balance, the linkage between generations would not exist at all if living generations prefer to avoid binding their successors with obligations to fulfil their expectations, if it allows them not to be bound by duties to their predecessors. In fact, the will to preserve the freedoms of the next generations against obsolete conceptions and values from the past and, more generally, against various forms of control of elder generations over younger ones, is not uncommon and is morally advisable. Its symmetrical element is the will of living people to preserve themselves against the obsolete conceptions and values of the past. In this search of a balance between lifetime-transcending interests and freedoms of each generation, there are multiple equilibria within society, without a dominant focus on one of them.

Returning to climate change, we may doubt the universal high intensity of the lifetime-transcending interests of living generations to preserve the climatic conditions experienced in the twentieth century when we see the difficulty of progressing towards a low-carbon economy and the collective unwillingness to control greenhouse gas (GHG) emissions at a sufficient pace. Thompson's insightful theory is grounded in an anthropological postulate that does not reflect historical and current experience. The theoretical foundations she gives to her idea of duties to the dead and the unborn are too fragile to bear all the duties and obligations she describes in terms of responsibility for and to the past and for and to the future.

5.3 JUSTICE BETWEEN DISTANT GENERATIONS: A CONTROVERSIAL CONCEPT

By taking the opposite approach to the one advocated by Brian Barry (1978), which aimed to liberate the norms of justice from any limitation resulting from consideration given to the circumstances in which the call for justice is made, I assert that the link between justice and situations should be given more attention. This is even truer regarding the issue of which relationship living generations may want to establish, albeit indirectly, with distant future generations. In fact, we need to radically question the relevance of the concept of justice to considering the relationship between distant generations, those which are at no time co-present in this world. After having been critical of Thompson's theory, do I mean that present generations are morally free from any concern for the joint long-term fate of humanity and the Earth? My answer is no, but I think this is not an issue of justice. We should change the conceptual basis mobilized to this effect.

Despite what Rawls himself thought, since his difference principle had to be understood dynamically by integrating the perspective of economic growth, thinking about the circumstances of justice raises questions about the relevance of the concept of justice to shaping the imaginary relations that living generations may have with distant future generations. Criticism can be formulated on several counts.

5.3.1 The Issue of Non-Existence

Rights and obligations

Distant generations do not exist in the present time; this is self-evident. Yet this fact raises considerable difficulties for any reasoning granting the idea of rights a primary ethical status, superior to the one recognized for well-being, for instance. Can we or should we grant rights to future generations, by which I mean effective rights in the present time of the society of the living, without waiting for them to come into the world? This is a subject of heated debate. Some authors (Wallack, 2006; Tremmel, 2006; Thompson, 2009; Gardiner, 2011) give a positive answer by associating two elements: (1) common intuition that we cannot ignore the fate of generations that are not yet born, and that we have obligations towards them; and (2) the postulate of a symmetrical relationship: 'obligations for some' means 'rights for others'. Thus, since intuition suggests that present generations have obligations towards future generations, this implies that the latter have rights over present generations. For some scholars, the link is then reinterpreted the other way around: it is because future generations have rights that are effective in the present time that present generations have obligations to them. Hence endless questioning about what is the first foundational term: the rights of future generations or the obligations of present generations to the former?

Point (2) is far from being unanimously accepted. Many authors support to varying degrees a two-way disjunction of the concepts: rights to people can be recognized, but they do not always create symmetrical positive obligations for their contemporaries, entailing only negative obligations to avoid being an obstacle to their fulfilment. This is what Hans Jonas (2000) called weak rights: you have the right to marry someone, but none of the members of society has the obligation to be your partner or to procure you a partner. There are also cases of moral obligations – for instance regarding animals – that do not have counterparts in the form of rights of other people (Gosseries, 2008). In the case of future generations, Dieter Birnbacher (1994, pp. 88–90) considered that present generations have obligations of concern regarding distant future generations, but denies that the latter, as such – in other words, unborn human beings – have rights in the present time. For person A to have a right over person B, it is necessary that: (1) A and B exist; (2) A has interests; (3) B has a moral duty towards A; and (4) A has the right to oblige B, alone or with others, to fulfil A's moral duty. Conditions (1), (2) and (4) are not met with future generations. In particular, Thompson's concept of lifetime-transcending interests cannot apply to the unborn.

Along these lines, several authors (Steiner, 1983; Beckerman and Pasek, 2001; Beckerman, 2006) have remarked that, since they have not yet come into existence, distant future generations cannot have rights: existence is a logical precondition for the faculty of a person to be granted rights. To have something – whether it is rights, interests, needs or utility – a living being must first exist. For example, the right to life, which is an essential human right, has no meaning for future generations not yet conceived. It is the same for other rights: the rights to health, safety, education, and so on. Never during their existence would the present generations be in a position to violate the rights of distant future generations. Let us replace the term 'rights' with 'obligations' in order to be convinced of this: can we say that future generations have, as such, present obligations towards the present generations?

Some authors (Feinberg, 1986; Meyer, 2004; Wallack, 2006; Gosseries, 2008) make a concession. Future generations do not have rights in the present time, but will have rights in the future that present generations can violate by their actions that will have effects extending into future time. In this respect, therefore, there is no obstacle to concluding that each generation has an obligation not to interfere with the rights of distant future generations. This was the case of past generations in the developed countries regarding the impact of climate change in the developing world, but they did not discharge their obligations (Davidson, 2008). Is this argument plausible? Two parables shed light on this discussion.

Two parables

Joel Feinberg (1986) used the following example: a cruel misanthrope intends to blow up a school and kill or injure as many children as possible. He hides a bomb in the toilets of a nursery school and sets the bomb to explode six years later. Everything goes as planned. Dozens of children aged five or younger are killed or injured. It was the evil action of the criminal six years earlier that triggered the catastrophe and the harm to the children. This action was initiated before the children were conceived and was nevertheless evil from the very outset, because it was going to violate the rights of human beings. By planting his bomb, he engaged a chain of causality that led to the damage it generated. The most common intuition, says Feinberg, leads us to consider that this criminal has committed a serious injustice in infringing the rights of unborn children from the moment the bomb was planted. This conclusion implies that the rights of children who have not yet been conceived, but who will come into the world in a future time, are sources of present obligations for present generations. The controversy therefore seems to be settled, and any discussion futile.

In fact, the interpretation to be given to the parable is not so clear. Firstly, note that the lives of the criminal and the children have a considerable element of temporal coexistence, even if the children are not yet conceived when the bomb is planted. For his act to succeed in accordance with his initial criminal intent, the criminal had to maintain that intent for all of the years during which the children attended the nursery school. If he had wanted to, he could have defused the bomb or told the authorities to do so. The final explosion was not inevitable, but contingent on the man maintaining his criminal will over time. This situation does not therefore differ from the most ordinary cases. It is the right to security of contemporary people that this criminal violates.

Once the children started to attend the school, it was the responsibility of the teachers and management team and, more generally, of the public authorities to ensure their security. If the bomb exploded, it was not only because the criminal initiated the process, but also because those in charge of the children's security failed to ensure the respect of their rights.

Considering the criminal's initial act, three modalities of moral judgement may apply, depending on whether the judgement focuses on the intention of the criminal, on the act itself or on its harmful consequences. None of these three impose the idea of an actual violation of the rights of people who do not exist. If we judge an act by its intentions, the intention to kill children 'in general' is reprehensible, irrespective of whether or not the children are already alive. If the moral focus is on the intrinsic quality of the act, the same judgement applies to the act of planting a bomb in a school, even if there may be special circumstances in which it would be legitimate, for example in times of war, if the school is occupied by an enemy who plans to exterminate an entire population (the lesser-evil theory).

If we judge the act according to its consequences alone, time regains its relevance, bringing its constraints of chronology and uncertainty. The judgement is divided into two parts: for six years minus one day, the act is considered totally harmless for the contemporaries and does not violate anyone's rights. When the bomb explodes on the last day, it destroys human lives and equipment and violates the associated rights. It is necessary to await this explosion so that the act can be judged as objectively bad in view of the destruction caused. When the violation of rights occurs, it concerns only living people at the time of the explosion. Moreover, as mentioned above, this event is contingent on the maintenance of the criminal's will and the failure of the school's supervision. The consequentialist judgement of the act therefore remains suspended until the moment the explosion occurs, if it does indeed occur. This judgement is not made as soon as the bomb is planted.

Let me modify one parameter of the parable: the time elapsing before the explosion, which I will increase to 120 years. Thus, the criminal will be dead at the time of the explosion and his criminal intention reveals a perverse 'lifetime-transcending interest'. According to the preceding analysis, his act will be considered reprehensible from the point of view of both the intention and the nature of the act, as in the previous configuration. But from a consequentialist perspective, his act will be considered totally harmless during the lifetime of the presumed 'criminal'. When the bomb explodes, it is an event that no longer has a living author. It no longer makes sense to want to condemn the bomber, if the goal is to inflict a punishment on him, even if the families of the victims are justified in lamenting the event. Meanwhile, these families have good reasons for turning against those who should have ensured the security and safety of their children (the school management, and so on). The death of the bomber makes the presence of the bomb an objective feature of the world, like a natural threat (the risk of earthquake, and so on) to future generations, that is an issue of safety.

The apparent evidence of the previous parable can blind one to the underlying moral reality. Limits can be illustrated by a second parable that steers intuition in the opposite direction: the intention and qualification of the action are morally different, but the nature of consequences remains similar. During the First World War, unexploded shells remain buried in the sand on a beach in northern France. A century later, children discover the shells and play with them. The shells then explode and the children are wounded or killed. Let us assume that those who fired these shells died in combat before the end of the war and are both victims and heroes of the war. Let us add that the soldiers knew the shells were dangerous and that some of them could be buried in the sand without immediately exploding, and could be a hidden danger for populations during and after the war, possibly for decades. Should those soldiers be considered as immoral and unjust because they violated the rights of children who were not born when the shells were fired? Who was responsible for the safety of the children? Should we incriminate the *poilus* for their acts a century earlier? What about the state having sovereignty over the beaches, which belong to the public domain, at the time of the children's accident? What about the children's parents, who have not exercised appropriate supervision? It is implausible and irrelevant to put the blame on the *poilus* for having violated the rights of children a century later. On the contrary, it is plausible and relevant to consider that both the parents and the public authorities in charge of the beaches are responsible for the violation of the children's rights and owe due compensation to their families. In this case, the intentions of the soldiers were not morally reprehensible, and their act was morally defensible during wartime, except for those who believe that any violence is always immoral. Regarding the nature of the consequences, there is no difference with the previous parable. For a consequences-based judgement, we should not be allowed to have different moral views on the cases described by both parables.

Limits to rights

Although most authors choose to distinguish moral and legal conceptions of rights in order to take account of the sometimes glaring shortcomings and inadequacies of the positive law of certain states, the possibility of translating the arguments in legal terms is a relevant test. The just institutions of a society are aimed at the devolution of rights to people, and in practice include impartial arbitration procedures between the conflicting claims made by the holders of these rights. The legal system can distinguish between the attribution of legal personality and the recognition of the status of an agent capable of asserting their rights on their own. For example, minors have legal personality but are not able to assert their rights on their own. The inability of future generations to assert their rights on their own is therefore not a theoretical obstacle, although it poses problems of practical organization. The main problem is elsewhere.

Proponents of the thesis of the present unjust violation of the rights of distant future generations base their argument on a specific way of shaping the concept of rights, which I call 'naturalist'. They assume continuity between the idea of a person's essential interests and that of their rights, which other people are required to respect and fulfil, without any social and institutional process being required to create those rights. In so doing, these authors do not pay enough attention to what is implied by the idea of rights and is not contained in that of interests. To have a right is to be able to impose a limitation on the freedom of others, either in that they must renounce certain actions, or that they must oblige themselves to take other actions. The difference between rights, interests and needs is that rights define what others should or should not do. Rights govern cooperation between living people whose lives, projects and choices intersect and interfere. The people who will come into the world at a future date will have rights, but these rights will be exercised within a perimeter limited to their contemporaries, those with whom they should cooperate and overcome disagreements. In a society with a certain degree of political and institutional stability, public institutions must at all times ensure respect for the rights of living people, with special attention to the rights to security and safety. It is this regime that applies to the schools and beaches. It should be anticipated by all living agents. There is no need to appeal to reasoning on the present violation of the future rights of human beings who do not yet exist. The same sort of regime operates with states and international institutions to avoid people becoming victims of climate change. It is a matter of contemporaries.

Another limitation relates to the content of the alleged rights. No one is entitled to ask for the moon. A right to something can be recognized only if that 'something' exists and is actually available in sufficient quality and quantity for the devolution of rights to make sense. Wilfred Beckerman (2006, p. 55) thus gives the following example: can we decide that future generations will have a right to see live dodos, the great bird species of Mauritius which died out at the end of the seventeenth century? Even if future generations had rights to the environment, they could not have rights over extinct species. In other words, the disappearance of species cannot be addressed in terms of a violation of the rights of future generations. Similarly, future generations cannot claim to enjoy a right to the climate that prevailed in 1960. They cannot therefore claim to suffer 'damage' or 'harm' because they have been deprived of this precise environmental state.

Finally, the situation described by the parables just mentioned does not establish a direct relationship between controversial acts (planting a bomb, and so on) and the conception of the children. They thus avoid the problem of 'non-identity' that will now be discussed in the context of the climate change problem. They were not appropriate metaphors for this intergenerational dimension of climate change.

5.3.2 The Non-Identity Problem

The problem of 'non-identity' was initially raised by Thomas Schwartz (1978) and developed at length by Derek Parfit (1984). It involves a logical paradox when the categories chosen to approach relationships between contemporary people are used in intergenerational relationships between distant generations. The paradox is as follows. To ethical individualism that acknowledges and valorizes the identity of each human being, individuals born in the future will never be in a position to blame previous generations for the state of the world they inherit at birth. Nor can they claim compensation or reparation for the harm they suffer. Had their predecessors behaved differently on a large scale, all mating relationships would have changed and these hypothetical future individuals would never have been born, being replaced by other individuals in terms of identity, quality and quantity. The paradox works for any theory of justice built on complaints or reproaches that future generations might make against previous generations because of the violation of their rights. The theory of reproach between distant generations is fraught with logical incoherence.

As Gosseries (2004b) showed in the context of French law on the case of a child born with a severe congenital disorder, the concept of harm becomes inappropriate when the feature we might be tempted to see as 'harm' is inseparable from the identity of the person concerned. In an alternative situation, the latter would not have lived at all. The prospect of a life without disability did not exist for that individual, assuming that medicine is powerless to cure the disability. Non-separability means that the alternative of 'the same person without the damage' does not exist, whereas the idea of damage is generally obtained from a comparison between several counterfactual states – 'what would have happened if' – that must correspond to real alternatives for the same person whose identity would not be affected. Consequently, such disabled persons are not harmed due to their disability since they are ontologically disabled, without an alternative that would preserve their personal identity. To put this idea in stark terms, a child born from the rape of his mother cannot claim to have been harmed by the rape, although his mother clearly was. Whatever the psychological difficulties of being such a child, there is no alternative for the child, whose identity is inseparable from this act. More generally, the expression 'to bring a child into the world' means this idea of intimacy between a human being and the world in which they are immersed. Ontologically, they form one single reality. This world historically changes from one generation to another, but this change cannot be said to be harm.

In the case of climate change, the populations that are suffering and will suffer the new climatic conditions attributed to the historical process of GHG accumulation in the atmosphere and of physical–chemical changes in the oceans cannot consider that they are suffering from unfair damage caused by previous generations that have disappeared. If the latter had behaved differently, those claiming to be victims would not have come into existence. For them, this new climate is not a disrupted climate: it is their climate.

These arguments show why the position defended by Caney (2008) considering global climate change as a violation of human rights is controversial, if not dubious, for future non-existing human populations, while remaining possibly sound for contemporary ones. Let us look at Caney's argument. He takes stock of modelled global assessments of the possible long-term impacts of climate change expressed in terms of increased malnutrition and health problems emerging against a counterfactual scenario assumed to be without climate change. He then claims that the human rights of future populations will be jeopardized by climate change. These populations can then legitimately claim compensation. Since the rights of future populations have been and will be impaired, past generations have committed a moral fault. This generates an obligation for their present heirs to compensate for this harm, independently of the fact that they themselves contribute to climate change or not.

Despite being attractive at first sight in terms of intuition, this argument is objectional in several respects. Firstly, what causes malnutrition and poor health? Do they directly originate in the climate? Certainly not. Climate change, as a global geophysical process, is never in itself a determinant of violations of people's rights. These evils come first from the structures of the society (the organization and productivity of production, the distribution of assets and income, access to basic services, and so on) in which affected populations live. Starvation is not a natural fate attributable to the behaviours of distant predecessors from other parts of the world. According to the domestic structures and policies of nation states, the same meteorological or physical event associated with climate change will produce very different impacts. For instance, tropical storm Matthew apparently caused the death of 1000 people in Haiti and none in Cuba in October 2016, although these locations have been hit by the storm under similar conditions. How should we consider this outcome in terms of who is liable?

Climate change does not directly impose on anyone either such inhuman living conditions that it would no longer be worth living, or living conditions that undermine the right to a decent life: at any moment, contemporaries living in decent conditions, if not affluent ones, have and will have the possibility and the responsibility to assist their fellow citizens who are exposed to a given manifestation of the climate threat. And those who would thus have an additional burden of solidarity towards contemporaries in need in the face of the climate threat, without themselves falling below the threshold of decent life, would not be in a position to blame past generations, since they themselves would be trapped by the non-identity problem.

Contrary to Visser 't Hooft (1999, p. 50), who claimed that he had never been able to understand the problem in the non-identity problem, Lukas Meyer (2004, 2008) and Moellendorf (2011) do not ignore it, but try to neutralize its effects. Their solution is to define harm or the violation of rights in an objective, comparative manner, rather than through counterfactual reasoning. Continuing with the concept of harm, Meyer believes that future people can be said to be harmed by actions that will take them below a certain prespecified threshold in terms of access to essential goods, even though harmful actions of previous generations are a necessary condition of their existence. To do so, he assumes two premises: (1) the problem raised belongs to the category of distributive justice, rather than to reparative justice; and (2) 'persons are under a general and universal duty to fairly secure a decent standard of well-being for all currently living and future people' (Meyer, 2004, p. 21). This solution exposes itself to two objections.

Firstly, how can we determine a universal well-being threshold across time in a non-arbitrary way? Is it possible to refer to basic needs without considering available technologies, institutions, access conditions, socially organized solidarity, and so on? These factors all change through space and time. At first glance, if the future state of the world is so disrupted and unbearable that it will no longer be worth living in it for all human beings, one could say that whatever the identity of future people, they will suffer major harm. However, in the context of climate change, this argument is clearly empirically irrelevant for the present time and many decades to come, and highly speculative thereafter: future contemporaries of people in need can always and should, if they are morally inspired, organize assistance to the less well-off in such a way that this extreme situation does not arise for them. Beyond that, on theoretical grounds, there is no rational line of reasoning by which an existing person can compare the utility of living with the utility of not having been born.

Secondly, since the issue raised concerns the impacts of climate change on living conditions, simply observing that some people are below the well-being threshold says nothing about the causal role of the climate factor in that outcome. In Meyer's definition, it is the imputation to climate change of some people falling below the well-being threshold that matters. To determine this fall, it is necessary to compare a state of well-being with climate change to a counterfactual one without climate change; this again raises the non-identity problem.

Leaving the 'damage' register for that of 'rights', Moellendorf considers his reasoning to be immune to the non-identity problem on the grounds that rights are independent of the identity of human subjects. Respecting the rights of people belonging to future generations should have been imposed in all cases on the previous generations, and the norm of equality of GHG emissions per capita would have drawn the dividing line between legitimate and wrongful use. Thus, for Moellendorf, previous generations had an obligation to compensate future generations for any violation of their rights. For example, they could have imposed a sufficient saving rate to finance future transfers of resources at the right level to achieve compensation.

However, Moellendorf's solution also fails to solve the problem. Firstly, he cannot avoid dwelling on the impossibility of defining *ex ante* the equal share of GHG emissions to which every person in every generation has a right. This is because of the huge uncertainty surrounding long-term demographic changes that depend on the free decisions of generations to come: one generation can never be sure what its fair share actually is (see Chapter 8). More central to this discussion is the following point: the equal rights of access to the atmosphere that Moellendorf attributes to the members of all subsequent generations would belong to an alternative world in which ownership of the atmosphere would have remained common and managed accordingly. This alternative world would not have experienced the economic growth that has been driven by a type of industrial growth making free and unlimited use of atmospheric and ocean capacities to absorb GHGs. On the other hand, by assumption, the world in which he believes previous generations should have taken measures to compensate future generations but did not, is the one in which appropriation of environmental commons has historically taken place, stimulating industrial development and economic growth, but causing climate change. In this historical world, inhabited by a population different from that of the alternative world, the idea of establishing equal rights per capita throughout history was not feasible: the extent of the rights granted to a population cannot escape the limits of the world that enabled the existence of that population. The granting of rights to a population cannot be based on assumptions implying that this population would not exist. Although rights are not as dependent on people's identities as harm, they are still part of a historical process linking demography, economic growth and GHG emissions.

Following Meyer or Moellendorf, a statement on the vertical relationship between past and future generations could go as follows: 'with climate change induced by generation Z, x per cent of the future population A could remain below the decent living threshold or below the level of equal rights to the commons'. It is important to underline that such a statement does not, in itself, qualify a climate injustice. This would require being able to define to what extent this outcome depends on the specific cause of climate change, since there are so many other factors that can explain why some people are unable to meet their basic needs in various regions of the world. To move forward, it would be necessary to envisage comparing two states of the vertical relationship, with and without climate change. We would arrive at a complementary statement such as: 'however, without climatic change, the percentage of population B below the threshold of decent living or not enjoying their equal rights to the commons could be y per cent'.

These statements mention population A and population B, since we are dealing with two different populations, in quality and quantity, in the two worlds compared. In the context of proposals by these two authors, we do not deal with marginal changes in climate states such as that caused by the emission of an additional tonne of CO_2. Thus, there is no option to compare the situation of the same population under two different sets of climatic conditions. Unfortunately, the horizontal comparison of the situations of these two populations A and B cannot be placed under the umbrella of the idea of justice, since this idea can only have meaning between people who coexist in the same world. It is impossible to escape the implications of the counterfactual rationale of non-identity by referring to objective norms of harm and rights.

To summarize, it is quite possible for a severe disruption of the climate system to significantly contribute to what we consider to be altering the future economic conditions of people's lives. This would not suffice to consider that climate change has directly violated the essential rights of future generations. The solution put forward by Meyer and Moellendorf is a false one in this regard. Non-existent people cannot be harmed by the behaviours of dead predecessors, since the concept of rights does not apply to them. Once they come into existence, the material condition in which some people are placed may be characterized in terms of disadvantages compared to the average condition of their contemporaries or to their hopes and dreams. Then, a society organized according to a principle of equal opportunity would certainly seek to mitigate or alleviate the disadvantages of its least well-off members (Roemer, 1998). However, these people cannot blame previous generations that are already dead for the suffering and other costs they must bear because of disadvantages. Within the framework just discussed, there is indeed no sense in the idea that later generations would have rights over previous generations that no longer exist.

5.3.3 Potential Ways Out

Can we envisage any ways around this negative conclusion? I see at least five options, with unequal consistency and value. They all imply painful costs or abandonments in terms of moral theory.

Coming back to utilitarianism

Because the attention given to the identity of people is at the root of the non-identity problem, a first option would be to abandon the moral individualism framework in favour of a doctrine that is indifferent to the identities and moral rights of human beings. This is the case of utilitarianism, as already seen. It does not provide a very solid basis for establishing an obligation of compensation for climate damage or the repayment of a historical climate debt to those who feel themselves to be the victims of climate change. It would be necessary to prove that such compensation and repayment would maximize welfare globally, taking account of the impact of international financial transfers on the respective economic growth of the regions of the world. The impact on growth would depend on the respective capacities of countries to allocate the financial resources they receive to an efficient development of productive uses. Most likely, the criterion of maximizing returns on investment would yield results different to those of proposals for fair allocation of emissions allowances made in the name of climate justice. If utilitarianism is associated with Paretianism, as proposed by most game theorists and economists, including Posner and Weisbach (2010, 2013), solutions would be even more distant from most moral proposals.

On theoretical grounds, utilitarianism may bring only an illusion of a solution. Ignoring issues of rights and identity, it routinely supports comparison of the welfare positions of abstract collectives of individuals in different scenarios. This could result in statements such as: 'people living in 2100 in a region X will have lower well-being, due to climate change, than people living in the same region without climate change, provided that those different people share the same preferences structure'. But this cannot be correctly translated as either 'people living in 2100 with climate change will suffer damage that they would avoid if no climate change occurred, and this is unfair', or 'people living in 2100 with climate change will suffer damage that people in a scenario without climate change will not experience, and this is unfair'. The last two judgements would not be acceptable for reasons already presented: (1) people will not be the same in the two scenarios; and (2) people in different scenarios do not coexist in the same world, and issues of justice and fairness can only be raised for coexisting people, in other words people who gain mutual advantages from their cooperation. It would only be through a total, but problematic, disconnection between the assessment of the well-being to be maximized and the identification of the people affected that utilitarianism could escape the non-identity problem.

Applying a generic ahistorical obligation to protect the climate

Another option is to seek to elaborate a moral argument, yet to be found, that every generation from 1850 onwards, including the oldest, had an obligation to take specific care to protect the climate for the benefit of all generations to come (Meyer, 2004; Davidson, 2008). Climate stability would have been placed at the top of the pyramid of superior goods, in the form of a categorical imperative, following the path advocated today by Piguet (2014). To avoid the non-identity problem, this obligation would be disconnected from any idea of harm imposed by the acts of previous generations on 'victims' at the individual or country scale, and hence from any compensation due to potential 'victims'.

As soon as the prospect of global climate dysfunction appeared in scientific circles in the nineteenth century, our predecessors would have had a moral obligation to adopt prudent behaviour leading them to refrain from using fossil energy sources, following a proposal of Marc Davidson (2008) that each generation should adopt precautionary behaviour in order to address the uncertain outcome of the non-identity problem. However, our predecessors did not act in this way and were therefore at fault. The current benefits of such wrongful acts would be illegitimate, which would provide a basis for the 'beneficiary pays' principle. Undue benefits should be handed over to the international community. The only difficulty with this reasoning – and it is a considerable one – is in establishing the premises for plausibly considering that this moral device was in place as early as 1850, which is not what Piguet claims.

In fact, this solution would be based on a historical misinterpretation assuming that in moral terms the ability of the atmosphere to absorb GHGs has been a resource common to all humankind since the beginnings of historical time. As with any natural resource, this function of the atmosphere emerged only when human technology and economic activity transformed a natural condition into a useful resource. Moreover, it was at the end of the twentieth century that it became a scarce resource due to the feared impacts of climate change. In other words, considering that the absorption of GHGs was a common resource in 1850 is the result of a retrospective illusion. It is blind to the contingency of the historical conditions for the emergence of the climate threat: the recent evolution of GHG emissions since 1988, when carbon dioxide (CO_2) reached 350 ppm, did not constitute a fatality or a fate that people in 1920, for example, could and should reasonably have anticipated. Not only did people at that time have no knowledge of the phenomenon, but the phenomenon itself had not been historically shaped. It was still possible that future energy systems would be based on diversified sources of energy, and that human demography would not take on the galloping pace we have eventually observed: in 1920, the world population stood at just 1.8 billion people. Futures without climate change were possible, which would have meant the absorption of GHGs was not to be a resource. Historical conditions have made the absorption of GHGs a resource that it was not initially.

Abandoning moral individualism for state ethics

One way to overcome the non-identity problem would be to abandon the postulate of moral individualism shared by most philosophers who have proposed their thinking on global climate justice, and to make states the basic agents of reasoning. Aside from a few historical accidents, states embody a lasting principle of historical continuity which, without suffering from the non-identity problem, ensures the preservation of freely subscribed rights and obligations, despite the vagaries of democratic life and changes in government. However, this solution is damaging to the demand for equal rights at the individual level (see Chapter 8). As the saying goes, you cannot have your cake and eat it too.

There are certainly good reasons for assigning responsibility to states: they are the basic agents of international law and international negotiations; the nature of pledges, policies and actions needed for mitigation and adaptation to climate change exceeds what can be expected from individual agents in terms of cooperation schemes and the volumes of resources to be mobilized (Page, 2012). Meanwhile, a moral problem must be considered. Is it fair to focus on states? There are two possibilities. The first is that states are considered as specific agents with specific modes of action, what we call policies. Their action takes place among the actions of other types of agents, including corporations, individuals, international organizations and non-governmental organizations, from which they are distinct and which could be held responsible for their own actions. Thus, having significant resources and freedom of action, states can be held responsible for their own actions, but only for their own. Overall responsibility should be distributed among the different actors, which assumes the availability of objective and accepted criteria to do so. The second possibility is that states are viewed as synthetic representatives of all agents forming a nation state. In this case, they take on responsibility for all the actions of all agents placed under their sovereign control. This solution requires a morally defensible doctrine of collective responsibility.[3]

Both options face moral arbitrariness issues (Caney, 2006). In the first case, the issue of fairness is posed by a procedure imputing all responsibility for environmental harm to one type of actor, the state, whose responsibility is only partially involved due to the active role of other types of agents. In the second case, the issue of fairness is raised because states do not live by themselves and necessarily redistribute their imputed responsibility to all agents under their legal supervision. If this redistribution is based on general fiscal rules, without considering the particular conditions in which individuals are placed regarding their contribution to climate change, it would not be difficult to find cases of clear unfairness and arbitrariness: inhabitants whose parents were recent immigrants from developing countries; people whose lifestyles imply a very low level of GHG emissions; and people whose emissions, even if rather high, are imposed by external disadvantages and the absence of reasonable opportunities, and so on.

The causally determined collective responsibility cannot be morally defensible unless two conditions are met: the willingness of individual members of nation states, and the actual opportunities they have to take part in the collective orientations and policy design that causally contributed to generating climate change and damage suffered by other countries (Miller, 2007). Without mobilizing the concept of odious debt,[4] which is relevant for extreme cases of rupture between a state and its people (Howse, 2007), in moral terms this democratic proviso is not met when we look back at the political conditions established in the developed countries over long periods since the mid-nineteenth century: many people, particularly women and foreigners, were not allowed to vote on the choice of political representatives or the approval of policy orientations. Moreover, if this condition of having the opportunity to participate in policy decisions is extended to the issue of damage linked to the past behaviour of dead people, living people have no possibility to influence the past and cannot be held responsible for what happened.

Miller admitted, however, that there was a need to distinguish moral responsibility from political responsibility, the conditions of engagement of the former being more restrictive than those of the latter. For the latter, he saw no objection to a people or a nation being considered as collectively responsible for harmful actions to other nations. He believed that such an imputation was legitimate in a democratic state, even for those who disagreed with what was done and acted to the extent of their resources to oppose directions of which they disapproved. As Miller wrote, it is in fact an inherent feature of democracy that the minority must accept the rules and orientations promoted and approved by the majority, with their attendant consequences.

Closing the time horizon on coexisting generations

Even if all net flows of anthropogenic GHG emissions suddenly amounted to zero, the physical impacts of climate change would nevertheless be felt for millennia. Over such time horizons, the cognitive foundations of an evaluation process based on the comparison of different costs and benefits accruing to distant generations are no longer guaranteed. Indeed, what would be the baseline enabling analysis by differences? How could we project a credible scenario of economic growth over such horizons, since damage is typically assessed in terms of the percentage loss of gross domestic product? What about demographics, or technical progress, and so on? Closing the time horizon is a methodological necessity if the exercise is designed to conserve both conceptual integrity and a link to reality. Even Thompson (2009) was aware that it was impossible to keep the language of duties and obligations either very far in the future or very far in the past. If these concepts are linked to the desire to maintain relationships within an intergenerational community (family, nation, and so on), they cease to apply at horizons when the existence of the specific community in question is no longer plausible. The same applies to the past. Thus, it is important to find a non-arbitrary way to introduce a finite horizon.

William Cline (1992) provided an empirical answer to this issue in his economic assessment of climate change. Based on the state of knowledge at the time on the carbon cycle, he conducted an evaluation of possible damage over a period of 300 years. He did so because, for this horizon, the greater part of atmospheric carbon emitted by present generations would have been recycled in oceanic and terrestrial carbon sinks. Implicitly, 300 years was the duration over which impacts on the atmosphere could be ascribed to the present generations. This horizon was therefore determined by the direct physical manifestation of human action and not by the objective features of global climate change (ocean acidification, sea level rise, rainfall regimes, ecosystem change, and so on), which will extend well beyond 300 years. In so doing, Cline gave rise to an intuition regarding another solution.

At one point, far in the future, the bulk of the climate change expected to occur will already have happened. The intensity of that change will depend on actions taken by the present and subsequent generations. Whatever it is, the people living at that future point will have radically adapted their lives and economies to the new global geophysical conditions. Their economy will have somehow merged into this new state of the planet. Under these conditions, it no longer makes sense to assess the new situation on Earth using the conceptual grid of damage defined by contrast with a counterfactual climate situation that will have vanished from any practical-historical set of possibilities, for instance the global climate existing in the mid-twentieth century. It is an intrinsic requirement of the cost–benefit approach that we compare alternatives to which we acknowledge a credible possibility of realization. The same applies to the moral assessment of harm and responsibilities.

To determine when this point in time occurs is an empirical matter, depending on the pace and severity of climate change and the types of impacts considered: changes in temperature regimes will come earlier than huge increases in ocean and sea levels. Thus, we can assume that there will be continuous changes to the global climate for a very long time to come, but humankind will then have been adapted to these climate dynamics as a permanent regime.

At this point, we may consider including a second closing point related to the 'non-identity' puzzle: the time horizon for decisions within which all coexisting generations are contained, but only those generations. To give a specific example, if the international community plans its actions towards a low-carbon economy until the year 2050, and we assume the standard lifetime of a human being in the twenty-first century to be 100 years, we should consider a time horizon of 2150 to analyse policy issues from various viewpoints, including fairness and justice.

Enforcing a moral and political standing for humankind as a collective person

This attempt to introduce a peaceful revolution into international law by building a moral standing for humankind as a collective person began with the qualifications of the 'common heritage of humankind' and the 'common concern of humankind' (Brown-Weiss, 1989; Godard, 1990; Kiss and Shelton, 2000; Shelton, 2009). Barry (1978, p. 204) believed that 'the value of continued human existence has to be a premise of other arguments that invoke the interests of actual (present and future) human beings'. Jonas (1984) developed the idea on the philosophical level by putting forward a new principle of responsibility for preserving the future existence of humankind as a living species accessing moral existence and moral responsibility. This future-oriented principle is not based on reciprocity between equal cooperating partners, as represented by contractarian theory. It is radically asymmetrical, non-reciprocal and unilateral by nature. Future generations cannot give anything to present generations in return for the care the latter may take in view of their well-being except, indirectly, the very opportunity of their own existence, which represents the reassuring prospect for previous generations that human history will extend well beyond their own lifetime. However, with this construct we quit the realm of justice and enter that of human desire to see the collective history of the species continue.

NOTES

1. Discussions of the discount rate abound, and were partly renewed by the release of the Stern Review on the *Economics of Climate Change* in October 2006 (Stern, 2007). See the book published by Resources for the Future (Portney and Weyant, 1999) and articles by Frederick et al. (2002), Weitzman (2007, 2009), Gollier (2013), Dasgupta (2007, 2008), Nordhaus (2007), Sterner and Persson (2008), Quiggin (2008), Caney (2009a), Posner and Weisbach (2010, Chapter 7), Fleurbaey and Zuber (2013) and Schaffer and Moellendorf (2014).
2. Giving the price of the numeraire of an economy, the discount factor at time t is necessarily unique for all the agents of this economy at that time. For this reason, I clearly cannot share the conclusion of Marc Fleurbaey and Stéphane Zuber (2013, p. 592) that: 'The key message of this article is that discount rates are really to be computed between individuals (person-to-person), which gives a great role to inequalities within and between generations'. From a moral point of view, this approach introduces considerable confusion regarding the respective responsibilities of previous generations and contemporaneous fellow citizens in addressing the issue of inequality.
3. For an extensive and thorough examination of the idea of national responsibility, which he supports without adhering to a cosmopolitanist viewpoint, see Miller (2007).
4. 'Odious debt' can be legally defended as a legitimate motive for refusing to repay debt when three conditions are met: the debt does not serve the national interest but the interest of the people at the head of state; the debt is contracted without the consent of the people; and the lender is fully aware of the previous points.

6. Historical responsibility for climate change

In the context of climate change policy, the term 'historical responsibility' refers to the proposal to define the obligations of a country based on its level of cumulative greenhouse gas (GHG) emissions since the beginning of the industrial era. More specifically, the emissions targeted are those produced between 1850 and 1990, the United Nations Framework Convention on Climate Change (UNFCCC) reference date. Although there is no intellectual necessity, this idea has often taken the form of a challenge to the developed countries whose past behaviour is said by some parties to violate the rights of the populations of the emerging and developing countries through excessive GHG emissions. This is because the latter countries would suffer damage from a changed climate and new obstacles to their development, while at the same time being deprived of some of their fair share of GHG emissions rights under a constrained intertemporal global carbon budget. This idea usually raises two issues: (1) the obligation to pay compensation for damage to 'victim' populations and countries; and (2) future penalization, in the form of a lower allocation of rights than the countries would have been entitled to, had they kept to their fair share since the early days of industrialization.

The moral and political plausibility of this thesis depends on adherence to postulates on the two terms 'responsibility' and 'history'. The first term applies to the conditions of the engagement of responsibility for actions contributing to climate change. The second concerns the validity of a retroactive imputation by which past behaviours are judged based on recently conceptualized norms.

6.1 THE MORAL CONDITIONS FOR CHALLENGING THE LIABILITY OF A PERSON

The foundations of moral thought on responsibility (accountability, liability) assume individual persons to be the subjects of responsibility. The understanding of the concept of fault attributable to a person in respect of damage suffered by another party presupposes that several conditions are met. The damage should be empirically constituted and affect protected rights. The causal link between the damage suffered by one person and the action of another person to whom this damage is to be attributed should be well established. The incriminated action has to be voluntary and free. The person who committed this action must have been aware of the likelihood of the harmful effects (Moellendorf, 2014, p. 165). The objective finding of a causal contribution is not sufficient, in moral theory, to establish obligations derived from liability. There are also subjective conditions relating to the authors of the action in question.

6.1.1 Different Nuances of Responsibility

There are different notions of responsibility involving varying degrees of linkage between observed damage and the authors of suspected actions. In his reflection on the idea of nations' responsibility regarding global justice, Miller (2007, pp. 86–104) introduced several useful distinctions. The first one contrasts agent-based responsibility (type 1) with the responsibility of remedying a situation or assisting another person, or 'remedial responsibility' (type 2). The two are articulated when it is the responsibility of the person who has committed a harmful act to another person to repair the damage done. In other cases, the two types of responsibility are separated for objective reasons, for example the death of agents who should have assumed type 1 responsibility, or for reasons relating to the organization of society, for instance with specialized public street cleaning services.

Type 1 responsibility is itself the subject of several distinctions. The less demanding notion of the subjective conditions to be met is pure 'causal responsibility', which considers human action as just one component among many that form the chains of causality. Here, all morally relevant subjective aspects are dismissed because the purpose is not to assign responsibility, but to understand what happened to produce a given outcome. Yet it is this notion of causal responsibility that proponents of the idea of 'historical responsibility' for climate change put forward in a moral and political judgement. Generally speaking, this type of responsibility is considered to be inappropriate or insufficient, both morally and practically, to imputing obligations. Random events in which agents are revealed to be physical contributors, to varying degrees, to the damage incurred by others are so numerous that agents would be constantly obliged to pay compensation to all kinds of people, even when their contribution did not exceed their legitimate rights or was imposed on them by the circumstances. Miller takes the example of a road accident occurring in the context of traffic congestion. Should all motorists be held responsible for the accident, and not just those who directly caused it, because they all contributed, however little, to traffic congestion?

With what Miller (2007, p. 88) calls 'outcome responsibility', causal links remain essential, but the purpose is different: it is a matter of gathering the elements enabling the imputation of responsibility, including the relevant subjective conditions of action. The consequences of action must be reasonably foreseeable, and such that a reasonable person with normal capacities could have avoided them by prudent action. The action must be free from coercion or any manipulation. When the causality chains are long and tortuous, Miller writes, responsibility dissipates. But outcome responsibility extends beyond the agent's own intentions, since outcomes may result from an omission or a lack of attention. Finally, this type of responsibility does not entail a moral judgement blaming the author to whom responsibility is attributed. In this sense, it remains objective.

Despite its intuitive appeal, under certain circumstances, this second notion embodies an overly broad and unlikely form of responsibility to legitimize a duty of compensation or reparation for negative outcomes for other people. For example, should the winner of a competition offer compensation to the losers because it was quite foreseeable for all competitors that the winner would take the prize and in so doing cause a loss – at least the initial stake – and disappointment to the other competitors? Another case is as follows: should a genius inventor, whose inventions are hugely successful with consumers but ruin the producers of competing products now made obsolete, compensate the losses of those unfortunate competitors? This would not be plausible. It is therefore appropriate to add other conditions to obtain a more relevant principle of responsibility.

The third concept, that of 'moral responsibility', is applicable where the agents who causally contributed to the creation of the damage acted wrongly by failing to comply with pre-existing moral obligations, including prudence, and thereby violated the rights of others. This view is more restrictive than the previous ones, since it is necessary to establish malicious intent, carelessness or negligence and a violation of the rights of others in order to mobilize it.

6.2 THE PREMISES OF A 'HISTORICAL RESPONSIBILITY' FOR CLIMATE CHANGE

According to Jonathan Pickering and Christian Barry (2012), historical responsibility for a climate debt is based on the following structure:

1. Countries are morally responsible for having emitted more GHGs than their fair share.
2. These responsibilities give birth to debts claimable by those to whom the debt is owed.
3. The responsibility of each country is determined by the magnitude of its cumulative emissions since the industrial revolution.
4. Climate debts are specifically owed by developed countries to developing ones.
5. Repayment should be made by paying compensations of adverse impacts and by allocating lower future shares of GHG emission rights.

Three additional premises are involved:

6. Damage due to climate change already exists and can be clearly distinguished from the impact of natural climate variability.
7. Its origin is to be found in GHG emissions prior to 1990.
8. Cumulative historical emissions are mainly attributable to the developed countries.

Premises 4 and 8 have been empirically refuted, since in the 2010s the balance of historical cumulative emissions is shared almost equally between the developed countries and the emerging and developing countries, as already noted. Premise 6 remains largely unresolved from a scientific point of view at the local or regional level, whilst being plausible at the level of global trends in the physical evolution of the planet. As the French climatologist Hervé Le Treut (2015) indicates:

> The amplitude of climatic changes linked to human activities has not yet exceeded that of the manifestations of natural variability. Variations in El Niño, monsoons, the Azores High, violent storms and prolonged droughts all bear the hallmark of predominantly natural dynamics, where chance also plays an important role, that GHG-induced global warming is disturbing . . . it is never possible to designate human activities as solely responsible for a given climatic event.

The other premises are examined in what follows.

6.2.1 A Nice Story by 'Uncle Shue'

To provide a specific intuition of the moral value of the idea of the historical responsibility of states, Shue (2009) made use of parables that he believed should be decisive in overcoming all reservations or criticisms. I now consider one of the most striking of these parables.

In order to cross a desert, one camel is chosen to carry the trunks of four travellers. However, the camel's real carrying capacity is limited to three trunks. When the fourth traveller puts his trunk on the camel, the animal falls down. Following Miller (2009), who adopted the same story for his own reasoning but with different conclusions, Shue asks who is responsible for this fall and for the journey being cancelled. Only the fourth traveller? The lesson Shue wants to impart is that there is no reason for the person using a common resource after others to be the only one to be blamed for the disruption of that resource. All users should be equally blamed or should be equally concerned by avoiding the overload.

Applying this lesson to climate change, Shue states that the fact that the developing countries are taking off and developing after the industrialized countries does not imply that they alone should bear the burden of resource limits. Those who have exceeded their equal 'rights' compatible with the overall carrying capacity should first compensate for this excess in order to leave a fair share available to the others; they should also compensate for the damage they have caused by exceeding the climate system's carrying capacity.

Does this parable provide a just metaphor for responsibility for climate change? Miller (2009, p. 132) did not think so: in the parable, the travellers are presented as being in a strictly symmetrical position regarding the use of the common resource that is the camel. However, this is not the case for the climate problem. Let us return to the narrative.

Firstly, people sharing a common goal (the destination of their journey), embarking on a joint venture (crossing the desert) and pooling an essential element without which the whole company would fail (the camel) in a common time (travellers travel together), are bound by a contract (the common rental of the camel) that gives each of them a right to the joint use of that resource. This right is not absolute; it is limited by the carrying capacity of the camel because the owner intends to recover his camel alive and well at the end of the journey. The travellers are therefore required to cooperate in order to manage the load properly to ensure they reach their destination and honour their contractual relationship with the owner of the camel. They form by contract a community of action within a unity of time and place.

Secondly, each traveller is deemed to have the same needs as the others. Thirdly, the short time elapsing between the different moments at which the luggage is loaded onto the camel's back does not alter the symmetrical position between the travellers, since the order of presentation is contingent; it also enables rapid corrective action to solve any problem of excess load.

Other lessons could be drawn from other hypotheses that are more appropriate to the climate case. For example, the first three travellers made their journey together and loaded the camel three days before the fourth traveller began his. They did not know that the camel was going to make a new journey three days later and would be so exhausted by this first journey that the mere fact of having to carry a new load would bring it to its knees. Perhaps this second journey had not yet been decided when the first three travellers made theirs; it might in fact never have happened, being a contingent event beyond the contractual links accepted by the three travellers. And if the camel had rested for eight days after the first trip, its strength would have returned, and none of the first three travellers would have been responsible for wrongdoing either to other travellers or to the owner. The only thing that can be said of the fourth traveller is that his contract with the owner was unfair because it was agreed between him and the owner that he was hiring a camel to make a specific journey, whereas the camel was unable to carry the expected load. If there is a liability to challenge, it is not vertical, linking the last traveller to the previous travellers, but horizontal, and concerns the contract with the owner. This is a rather different conclusion to the one given by Shue.

Although superficially related to the camel parable narrated by Shue, the climate problem is in fact very different. It results from a historical GHG accumulation process involving two stages. In the first stage, the GHGs that accumulate in the atmosphere remain compatible with the targeted safe cap of GHG atmospheric concentration. Thus, within these limits, changes in the terrestrial–ocean–atmosphere system do not result in dangerous modifications to the climate regime beyond anything experienced since the beginnings of human civilization. The damage associated with different climatic or geophysical events remains within the limits of what humankind has experienced until then because of natural variability. Once this threshold is crossed, a second stage begins, in which the climate machine begins to be unbalanced and deviates towards other long-term equilibriums.

The solution put forward by 'Uncle Shue' attributes to the first drop deposited in the 'GHG bottle' some 150 years ago, the same moral responsibility for the consequences of the overflow as the drop that actually caused the bottle to overflow from 1988 onwards. The issue is presented as if all users of all generations had gathered in front of the empty GHG bottle and had together defined a plan of egalitarian use to keep GHG emissions within the capacity of the bottle. In fact, the first users did not even know that there was a bottle or that it had a limited capacity. Users much later understood that there was perhaps a bottle, but they knew neither its capacity nor the number of users expected, since this number had not yet been defined in reality, nor had the extent of users' needs. All users were unable to define their fair share in a fictitious intertemporal distribution, since the constraints of the game were undetermined. In other words, the conditions for challenging 'moral responsibility' and even 'outcome responsibility', according to Miller's definitions, are not met, at least for a very large part of the historical period considered (1850–1988).

6.3 FACTORS OF EXEMPTION FROM HISTORICAL RESPONSIBILITY

One foundation of moral responsibility is that everyone should bear the consequences of their free and informed actions inasmuch as they result in a violation of the rights of others. In this regard, the idea of the historical responsibility of human beings for climate change encounters limits due both to the reality of things and to the state of knowledge.

6.3.1 Ignorance?

A first assertion is that past generations are not responsible because they did not know about the global climate change phenomenon and were unaware of its adverse impacts (Bell, 2011). When the first scientists spoke of the possibility of climate warming induced by carbon dioxide (CO_2) emissions, this was to emphasize the benefits for Northern countries such as Sweden. Scientific research in the 1970s still highlighted the concern that the climate would shift towards colder temperatures that would threaten agriculture and geopolitical security. GHG emissions were perceived by some as a beneficial means of mitigating this threat. In the 1990s, scientific and social controversies about the reality of climate change and its causes were still widespread. These continued for 20 years, peaking in 2009–10 around the Copenhagen Conference in December 2009 (Zaccaï et al., 2012).

Thus, the generations that died prior to 1970 could legitimately be unaware that the use of fossil fuels and changes in land use could have a huge negative impact on the climate. After 1970, people could reasonably have heard of a potential risk of global climate change, but not know whether this would entail warming or cooling and whether the causes were natural cycles or human GHG emissions. It was then impossible to paint a picture, even an approximate one, of the potential damage incurred in different parts of the world. The 1980s generated progressive awareness at the interface between the scientific and political worlds, but this did not penetrate public opinion as a real threat. The Intergovernmental Panel on Climate Change (IPCC) was established in 1989 and submitted its first report in 1990. That year was the pivotal year when knowledge about the risk of climate change was sufficiently disseminated throughout the world to ensure that caution could be morally expected. The content of this caution remained to be determined, however.

There are few challenges to the elements that have just been recalled, but the debate has focused on the following: does ignorance prior to 1990 exempt current populations from the presumed 'historical responsibility' of their country? Academic researchers among the most committed to defending a cosmopolitanist approach to the climate problem (Shue, 1999, 2009, 2014; Neumayer, 2000; Gardiner, 2011; Vanderheiden, 2008) have argued that ignorance is sufficient to exonerate guilt for moral misconduct, but not to erase the obligation to compensate for the cost of damage suffered by the victims of climate change. The argument refers to common usage, common sense or, more precisely, the concept of 'strict liability', without fault, acknowledged by several legal systems under certain circumstances.

The 'strict liability' regime has been established in the USA for activities that, though necessary or justified, are abnormally dangerous for human health, and are perceived as such *ex ante*. The aim of this norm is to generate a high level of vigilance at all times on the part of the human agency in charge of these activities. It is intrinsic to the use of this concept that agents responsible for particularly hazardous activities know in advance that they will bear the cost of damage in case the risk is realized, whatever the circumstances (Moellendorf, 2014). However, the unknown nature of the risk at the time activities are engaged provides legitimate grounds for exemption from liability. The European regime for defective products also provides for an exemption from liability for what is known as 'development risk', in other words the subsequent discovery, through use, of a form of hazard unknown at the time a product was launched.

Mobilizing this concept retroactively, for the actions of people who were unaware that they would be subject to this regime of responsibility, would be a serious illegitimate deviation (Schüssler, 2011). Strictly speaking, this concept of responsibility could only be used for GHG emissions after the date at which its application would be decided at the international level.

From a moral viewpoint, Derek Bell (2011) proposed an intermediate 'limited liability' solution based on the distinction between the moral qualification of the acts (the emissions of GHGs between 1850 and 1990) and the moral judgement on the behaviour of those who committed these acts: are they blameworthy or not? Bell argued that it was possible to reply negatively to the second question, while considering that the acts committed were wrong, since they involved damage borne by others. It would be 'wrongdoing without fault', in a way. Because of the wrongful nature of the actions, a moral agent to whom it is revealed why and how he has done wrong in the past should regret his actions, should not want to keep the profits he gained from these activities and should want to compensate the victims to the extent of the benefits he unduly obtained. The same should apply to his heirs.

The controversy over the excuse of ignorance is thus not closed, and is divided between a legal concept with restrictive conditions of application and an extensive moral conception based on the possibility of separating judgements on acts themselves from judgements on the agents who committed these acts. However, the morally 'wrongful' nature of historical GHG emissions remains to be established. Likewise, the obligations of the beneficiaries who are, by hypothesis, the heirs of those who have innocently committed the acts deemed to be morally wrong, remain to be established.

6.3.2 The Present Generation's Lack of Control over Past Emissions?

It is generally accepted that it is arbitrary to attribute responsibility to people for acts over which they have no control, especially for actions committed by others, unless specific attributions in society give them this remedial responsibility. But, according to the thesis of historical responsibility, the present generations should assume the consequences of past actions over which they have had no practical control. This deserves a discussion.

For an analysis of responsibility, with a life expectancy of 90 years and a restriction of the target to only adults, all GHG emissions prior to 1945 certainly fall into the 'beyond control' category. These emissions account for 36 per cent of atmospheric CO_2 accumulation between 1850 and 1990 (European Environment Agency, 2013). Moreover, population dynamics in the twentieth century have meant that more than four-fifths of the population of the developed countries were born after 1955 and became adults at the earliest in 1973. These people should not be concerned by historical responsibility for emissions prior to 1973. Consequently, they could be held responsible for only a third of the emissions produced between 1850 and 1990 (atmospheric CO_2 concentrations were 328 ppm in 1973, 285 in 1850, and 354 in 1990). Given these exemptions, the historical responsibility of the developed countries could be only partially applied in the hypothetical case in which it is considered as a valid foundation. A considerable proportion of climate damage would not be compensated at all.

6.3.3 Counter-Argument: the 'Beneficiary Pays' Principle

Although they are not directly responsible, people living today enjoy benefits and wealth that are the result of presumed reprehensible or unjust activities conducted in the past in the country in which they live. As the beneficiaries of wrongful or unjust acts, they have a duty to repair the damage done, or at least to retrocede undue benefits. This argument has resulted in what several authors (Gosseries, 2004a; Page, 2012; Baatz, 2013) call the 'beneficiary pays' principle (BPP). Robert Huseby (2015, p. 210) gives the following definition:

> If an agent A benefits from an action x that she did not perform, and an agent B is harmed by x, then, if B did not perform x, A, in virtue of benefiting from x, owes B compensation (which is not to exceed A's gain from x), insofar as no other agent has stronger duties to compensate B.

The present inhabitants of the developed countries should consequently return to victims the undue benefits acquired since 1850 because of the economic growth enabled by GHG emissions.

Let me begin by mentioning, without discussing them, the empirical objections to the previous allegation:

1. The size of the benefits involved should be in the right proportions. The collective wealth accumulated by countries for a century and a half is not primarily the result of GHG emissions, although past GHG emissions have been one of the many by-products of the path of economic growth that has been followed; the quality of economic institutions and political stability weigh more heavily.
2. Little is left of the legacy of industrialization prior to 1945. With technological innovation revolutionizing production patterns, the benefits of a generation of industrial investment are important in the medium term, but not in the very long term. For many of the former industrialized regions, the discovery of the limitations of the industrial promise was a bitter pill; for them, what remains of decades of investment in mining and heavily energy-intensive industries?
3. There is no reason why historical responsibility should be limited to the inhabitants of the developed countries. The other countries weigh significantly in total cumulative emissions up to 1990. Moreover, the industrialization of the developed countries has enabled flows of technological invention and innovation and huge advancement of knowledge from which all the countries of the world have benefited in most fields (industry, agriculture, medical care, and so on). This is now enabling the developing countries to avoid taking the same steps as countries that experienced early industrialization (Goldemberg and Guardabassi, 2012). Does this contribution to the global progress of humankind not compensate for the negative impacts of GHG emissions?[1]

The theoretical discussion has brought out various arguments. Edward Page (2012) admits the indefensible nature of the thesis of the wrongfulness of historical emissions: there are no precise causal links between the emissions from one country and the harmful climatic events suffered by another; no emissions from any country can be considered wrongful prior to 1950, being insufficient to generate a dangerous interference with the climate system; the dual conditions of the foreseeability and control of the harmful consequences of those emissions are not met. However, Page argues that past GHG emissions were the source of 'unjust enrichment'. He proposes three possible versions of the factor characterizing injustice. None are as convincing as he believes.

In the first version, these GHG emissions have enriched some people and reduced the well-being of others, although the latter have contributed very little to climate change, unlike the former. The latter having been disadvantaged, Page says it is sufficient to describe as unfair the actions that caused the damage. The second version is as follows: because of the atmospheric accumulation of GHGs, the developed countries have created new obstacles to the development of the other countries by depriving them of the opportunity the developed countries have had to industrialize without having to worry about the constraints of GHG emissions, and this is unfair. According to the third version, the ability of the atmosphere to absorb GHG emissions, along with that of the majority of ocean carbon sinks, is by nature the common property of humankind. This entails an equal right of use for everyone on Earth. Historical emissions from the developed countries having clearly exceeded these countries' user rights at the expense of the other countries, injustice is characterized.

6.3.4 Objections to the 'Beneficiary Pays' Principle

Taking the opposite view to Page, several authors have rejected the general value of this 'beneficiary pays' principle when it refers to damage *lato sensu* and not to an injustice originating in a fault. This is the case of Miller (2009), Caney (2010), Ewan Kingston (2014) and Huseby (2015). Most of their arguments relate to the first interpretation proposed by Page, and invalidate his proposition. The mere fact of receiving involuntarily and without fault a benefit from the action of a third party does not create an obligation to return this benefit to people who find themselves disadvantaged by this action. Many parables are used to mobilize intuition to confirm this conclusion.

Paul loves Susan desperately, but Susan is more sensitive to the charm of Peter, with whom she decides to go and live. Paul's suffering is intense. Should Peter offer compensation to Paul? Or should Susan do it? This parable suggests that it is not the harm or loss of well-being inflicted upon another party that can create the obligation to compensate the 'victim', but the violation of a protected right.

John is walking in a street in his neighbourhood. From an open window, he can hear a recording of Jimi Hendrix's one and only concert at Woodstock. John takes great delight and stops for a moment to listen. Peter comes along, plagues John about this 'music for heathens' and hurries on. The same action generates an unsolicited benefit for John and an imposed displeasure, if not suffering for Peter. Should John compensate Peter to the level of his pleasure?

Peter owns a house on a hillside, and decides to undertake work to build a cellar. Unfortunately, this work causes a landslide that partially destroys Paul's house below, but also uncovers a layer of gold on Jason's land, making him suddenly as happy as if he had won the lottery. Following these events, Peter dies of a heart attack. In this situation, does Jason have a moral obligation to compensate for the losses suffered by Paul? And would Paul be morally right to insist that Jason pays him compensation equal to the damage he has suffered? It would undoubtedly be morally praiseworthy if Jason did so, at least partially – by virtue of altruism and care – but would he morally offend justice by not doing so?

A rapid analysis might lead one to believe, wrongly in my view, that Jason is the beneficiary of an injustice. In fact, Jason's gain does not arise from the fact that Paul's house was destroyed. Jason's gain and Paul's loss have a common origin, but it is not Paul's loss that benefits Jason. There is no causal link between the gain of one and the loss of the other. The situation would be quite different if Jason, who had been watching Paul's land for a long time, took advantage of the opportunity provided by the destruction of Paul's house to purchase this land at a low price.

In the case of climate change, the present generations in the developed countries are not the beneficiaries of excessive GHG emissions – they are also the victims of climate change – but of products and services derived from the use of fossil energy sources and invested in modern production facilities. The common cause connected by the energy technology prevailing in the past since the nineteenth century does not create a causal link between the benefits for some and the damage to others. It would be different if present generations in the developed countries gained positive benefits from the climate damage suffered by the developing countries, for instance by buying land at a low price from people reduced to a strategy of immediate survival.

Going further in the critical exploration of the BPP, Caney (2006, pp. 474–5) mobilized the 'non-identity' problem (see section 5.3.2 of Chapter 5) to argue that members of present generations could not be the beneficiaries of an alleged injustice committed by their dead predecessors, because if the actions qualified as injustices had not been committed, these generations would never have come into existence. They cannot therefore truly regret that these actions have been committed, just as the alleged victims cannot claim to suffer harm attributable to the unjust behaviour of the predecessors of present generations in the North, since their coming into the world depended on this behaviour. For all parties, these past behaviours have meant that they exist; they and not others.[2] This is the nature of the world.

One of the fallback solutions of proponents of the BPP has been to abandon the framework of reparative justice and switch to distributive justice, as Meyer (2004) already had for the general case. For instance, Christian Baatz (2013) puts his argument in the perspective of equal opportunity. In relation to a supposedly fair initial situation, the changes induced by GHG emissions unjustifiably increase the gap between those who benefit from the positive aspects and those who suffer from the negative impacts. This distributively unjust shift would therefore call for measures to restore equality of opportunity between the affected parties. The BPP would provide for this restoration and would be justified for this reason.

This argument does not satisfy Huseby (2015), because this form of correction of a presumed arbitrary inequality generates another more global inequality: those victims who are wronged by acts that did not generate the smallest benefit for third parties and were committed by agents who are no longer in a position to repair them, would be left without compensation. However, the generalization of the BPP – any agent benefiting indirectly from an injustice should compensate any agent suffering from an injustice, even if there is no causal link between them – would not make it more defensible. It is up to society as a whole, and not only to the supposed beneficiaries of an injustice, to take responsibility for policies capable of restoring an equal opportunity endangered by certain actions. In the case of climate change, from a distributive norm of equal opportunity, the victims of climate damage should be compensated, but it is the whole international society that should be the source of compensation, not the supposed beneficiaries of past excess emissions.

The second interpretation of the BPP proposed by Page is not more convincing. In addition to the fact that a comparison between distant historical periods (from the nineteenth to the twenty-first century) cannot easily lead to a qualification in terms of injustice – per capita living standards were much lower in the past than they are in 2016, so does it make sense to see this as an injustice? – Jose Goldemberg and Patricia Guardabassi (2012, 2015) refuted this thesis by stressing the fact that the emerging and developing countries now have access to technologies that were not available to the old industrial countries: the asymmetry is dual and reversed. The premise of the third interpretation will be discussed and refuted in Chapter 7. It meets other obstacles too.

6.3.5 The Trap of Retrospective Illusion

Most of the arguments discussed up to now assume that past generations have committed an injustice due to excess emissions of GHGs. Arguments about ignorance, lack of control and the BPP were put forward under this assumption. But this idea has not been tested and challenged in itself. It is time to proceed and to show that it is the outcome of a 'retrospective illusion'. This illusion consists in interpreting history as a deterministic sequence that should necessarily lead to the dated result known to the analyst-interpreter and that could be perfectly forecast by past generations. Then, the known result permeates the whole historical sequence as if it were already inevitably present at different historical stages rather than being just one possible outcome among others. So, past actors in history are attributed knowledge or an obligation of knowledge regarding the historical outcome, which could only be gained at the end of history.

When the historical result is damage, the retrospective judgement marked by the illusion will attribute this damage to all previous stages, as if the wrongdoing were already inherent in them. To paraphrase a famous formula, the 'end qualifies the means'. In fact, the illusion corrupts two dimensions of judgement: the relation to knowledge and the relation to reality.

In the case of climate change, not only did past generations not know about the prospect of this problem until the 1980s or 1990s (ignorance objections), but historical emissions prior to these years were not causally responsible for any dangerous interference with the climate system. Alongside the international negotiations held since the Copenhagen Conference (2009), there has been a movement by scientists (Hansen et al., 2013) and non-governmental organizations calling on the international community to establish a long-term concentration target of no more than 350 ppm of CO_2. This value is far more restrictive than those commonly discussed and recommended on the basis of the objective of a maximum increase of 2°C in temperature: around 440–450 ppm. Its advocates present 350 ppm as the level of concentration in line with the objective set under the UNFCCC. However, this level, which is certainly not a level that would have no physical impact, was not reached until 1988. Previous emissions were therefore compatible with the maintenance of the type of climate known to humankind since the beginning of civilization.

For an analysis of the causes of climate damage, conceived as abnormal and dangerous disturbances to the climate system, only emissions after 1988 should count: added to the existing stock of GHGs in the atmosphere and oceans, these recent emissions have tipped the terrestrial–ocean–atmosphere system over into a state of dangerous disorder. Here, we again find the metaphor of the GHG bottle that eventually overflows.

As already noted, the recent evolution of GHG emissions since 1988 was not an inevitability that people in 1920, for example, could and should reasonably have anticipated. The objection here is not that these people had no knowledge of the phenomenon, but that the phenomenon itself had not been historically formed. Very different futures were possible concerning the economy, the energy technology and demography. Scenarios other than the one that actually occurred were still possible, which would have led to a future without climate disruption. In retrospect, but without illusion, climate change in these past times had no reality. It was only a contingent possibility. Its realization would depend on the free choices of the subsequent generations, especially those that have steered economic and technological development since 1970, when the energy transition to energy efficiency and low-carbon energy sources would have had to be prepared. If people in the early twentieth century were not causally responsible for the problem, their heirs have no special obligation to repair or compensate for anything, or to relinquish undue benefits.

6.4 TWO SORTS OF DEBT

How should we understand the fact that the issue of historical responsibility has been so important in climate debates since the 1990s? I propose seeing in this construction the effect of an approach seeking to mobilize a conceptual innovation of modernity – contractual debt – for the support of older, even archaic conceptions of the idea of debt. The retroactivity over two centuries of rights and norms invented in the twenty-first century is based on the confusion of two different conceptions of what debt means.

The first type (type 1), which is strongly inspired by the domination of traditional religions over the social order, makes debt a constituent element of the human subject from birth: the 'debt of life' enshrines the essential dependence of individuals on entities to whom they owe their life: their parents, clan, society, gods or God, and so on. As such, people are subject from birth to a set of obligations and loyalties from which they cannot morally escape, not to mention social sanctions for transgressions. They must constantly mark their gratitude by all sorts of rituals and practices. This ontological debt is constituted outside of any legal framework or any voluntary contract, since it precedes all of them. It is also a debt that can never be paid; it is inextinguishable. Overall, this 'type 1' debt institutes a human relationship characterized by three attributes: it is asymmetric, pre-contractual and without term.

When this pre-modern conception of debt is projected into modern society, it is at best reinterpreted as the source of recognition of some of the dependency and contingency relationships that mark the lives of all people. What the human subject will then do depends on their own decision by which they delimit the responsibility they choose to assume. Refusing this 'debt of life' as the vector of the domination of family or social surroundings on their life, the modern subject must determine which of the loyalties and fidelities they decide to assume and which they intend to abandon or recuse. Still pre-contractual and asymmetrical, but attenuated and open to free choice, this reconfigured 'type 1a' debt assumes the ideas of the liberation and autonomy of the person from the relational fabric that enabled their constitution. The asymmetry of freely chosen sovereign responsibility is then the counterpart response to the asymmetry inherent in pre-contractual debt.

By making the concept of contract central to sociability, in the field of economics and commerce, but also in the reflection on political society conceived as being instituted by a social contract, modernity offered a means of promoting individual autonomy by disengaging it from the domination of an archaic conception of an inextinguishable debt that encloses individuals and peoples in the dense network of imposed loyalties. This contract gives rise to a 'type 2' debt that is established in return for services rendered and defined benefits. It introduces the principles of symmetry and delimitation. Here, debt is limited in purpose, value and maturity. It is refundable and must be repaid. But once the contract is executed and paid, the contracting parties are released from any special obligation. This 'type 2' debt necessarily proceeds either from a specific, explicit prior contract, or from general but precise social obligations resulting from the actual rules of civil and criminal law, which everyone is supposed to know. It establishes a symmetrical, contractual and limited relationship both in its content and in its duration. The two types of debt are thus constructed on completely opposite grounds.

Everyone can have views on the balances between the two types of debt in a specific historical society. But what is unacceptable is confusing them, or treating a 'type 1' debt according to a 'type 2' format. The conceptual approach that inspires the ideas of 'climate debt' consists in attempting to dress an archaic, pre-contractual conception of an ontological debt, here transposed to the geopolitical history of the world, in the features and content specific to 'type 2', those of a debt either derived from a contractual obligation or legally constituted, one which must be repaid by debtors to a creditor. This operation is the vector of a conceptual regression against an essential component of modernity. It manifests an archaic refusal of historical time, with its irreversibility, which marks the succession of generations who never have to live in the same world as their predecessors, for better or for worse.

6.5 CONCLUSION

Let me summarize the discussion. There are two major strategies giving substance to the idea of 'historical responsibility'. One shares the moral universe of an individualistic ethic and claims to dispose of all considerations that are morally arbitrary from the point of view of individual people. The other admits the idea of collective responsibility and is aimed at attributing accountability and liability to states, taken as moral and political agents. The former mobilizes a causal responsibility approach to characterize the excess per capita GHG emissions of the developed countries as a sufficient and well-established moral foundation for obligations of compensation and reparation. The latter would give more importance to the absolute level of a country's emissions and to collective cost imputation criteria, such as equalizing the burden of different generations in the percentage of gross domestic product.

To bring into play the idea of historical responsibility prior to 1990, advocates of an individualistic ethic should have, firstly, proved the existence of faults, injustices and damage resulting from the violation of protected rights; and secondly, found procedures for individualizing imputations within each society, without stopping at the level of states. On this latter point, proposals have remained extremely rare. In fact, this approach to responsibility has encountered several classical obstacles (ignorance, lack of control), but also the non-identity problem and the unacceptable bias of retrospective illusion. Thus, it was impossible for its proponents to establish convincingly, from a moral point of view, how emissions prior to 1988 were the source of specific obligations incumbent upon the present inhabitants of the developed countries.

The second approach, which is centred on the collective responsibility of states, is not affected by the non-identity problem because of the postulate of the historical continuity of these political entities. The balance of cumulative emissions since 1850 reveals a causal contribution divided equally between the developed North and the emerging and developing South. However, the legitimate ignorance of people prior to 1970 about the possible threat of climate change and, above all, the fact that atmospheric CO_2

concentration remained below the level of 350 ppm until 1988, dismiss the 'beneficiary pays' principle regarding inherited obligations of compensation for the contemporary victims of climate change. Moreover, they also dismiss any allocation scheme of future rights and obligations to states penalizing their historical responsibility prior to that date.

Ideas of responsibility for harm imposed by climate damage retain their validity for emissions produced after 1988–90. Conditions regarding knowledge of the threat and control of its sources are indeed met. For example, Miller (2009) and Moellendorf (2014) proposed attributing the costs of damage and additional obstacles to development caused by emissions after 1990 either in proportion to these GHG emissions, in Miller's case, or in proportion to the share of these emissions that exceeds subsistence emissions, in Moellendorf's case. In their view, this imputation should apply to all countries, whether developed, emerging or developing, except for the least-developed countries according to Miller. These proposals do not consider the compliance or non-compliance with commitments made under the Kyoto Protocol to be a relevant consideration, which is an issue for international justice.

NOTES

1. This argument is strongly disputed by moralists. As Davidson (2008, p. 477) puts it: 'the creator of external costs and non-bargained external benefits cannot cancel them out one against the other'.
2. On this argument that living people cannot truly regret past injustices, since their very existence depends on past generations having behaved wrongly, see the notable paper by Thompson (2000).

7. The equitable sharing of a carbon budget

For the future, avoiding dangerous anthropogenic interference with the global climate system implies introducing severe limits on net global emissions. What would be a fair worldwide distribution of the corresponding caps? How can countries share the efforts to be made by them in the near and distant future? The United Nations Framework Convention on Climate Change (UNFCCC) lays down the principles of equity, common but differentiated responsibilities and capabilities. The scope of criteria proposed in the literature is broader. We find equal per capita emissions rights, equal emissions abatement burdens, equal access between countries, 'historical responsibility', capacity to pay, the 'beneficiary pays' principle (BPP), a lexicographical prioritarian criterion for subsistence emissions over comfort and luxury emissions, and acquired or customary rights at a reference date (Grasso, 2012). In many proposals, several criteria are combined.

Regarding principles, there are four main inspirations for organizing the sharing of rights and obligations: ensuring respect for pre-existing rights; obtaining commitments in proportion to available means (capabilities); accountability for the creation of the problem (cumulative emissions); and taking account of different circumstances specific to each country, such as inequalities of income, wealth and exposure of populations to climate risk (distributive equity), to adapt the automatic application of the main criteria. The last three principles correspond to the three enshrined in the UNFCCC, which says nothing about the first, even though it is very present in the academic debate.

7.1 THE ISSUE OF RIGHTS INVOLVED IN CLIMATE CHANGE

The mobilization of the issue of rights leads in different directions. The first, (A), finds its basis in two intellectual operations: (1) postulating in a non-metaphorical way that the Earth is the common property of humankind; and (2) affirming that this common property necessarily entails the individual equality of access rights to the common natural resources that the Earth contains. The second, (B), is not concerned with ensuring equality of access to natural resources, but may be inspired by two different central values: a 'sufficiency' or 'prioritarian' perspective (B'), and a customary approach, be it libertarian or communitarian, to the conditions for the legitimate appropriation of natural resources (B''). The first, (B'), claims the existence of moral rights of people over the elements of their living environment upon which their fundamental interests depend, while the second, (B''), considers the customary origins of usage rights to the local environment, in particular in the form of greenhouse gas (GHG) emissions.

In perspective (A), several experts from developing countries and non-governmental organizations gravitating around the climate negotiations have proposed methods for calculating multi-decadal carbon budgets per country based on the principle of individual equality of rights, integrating the past since 1850 and the future until 2100. This type of calculation brings out a huge carbon debt of the developed countries, to be deducted from their future rights. For instance, Martin Khor (2012, pp. 1–2), director of the South Centre, a think-tank created to support the positions of developing countries in international negotiations, claimed that the fair share of industrialized countries was 310 $GtCO_2$ between 1850 and 2008, and their overuse was 568 $GtCO_2$. If the average unit value of one avoided tonne of carbon dioxide (CO_2) is $40, this means a climate debt of $22 720 billion. For a 2010–50 carbon budget of 750 $GtCO_2$, the fair share of industrialized countries is 120 $GtCO_2$. With restitution of the carbon debt, their allocation should be a negative budget of 448 $GtCO_2$.

Caney (2005b, pp. 767–8) seems to have adopted the second perspective (version B'). Indeed, he concludes his critical examination of the proposed solutions by advancing the following principle: 'Persons have the human right not to suffer from the disadvantages generated by global climate change'. This principle is also supported by Shue (2011), while Vanderheiden (2008) advocates the recognition of a 'right to climate stability'. Noting that a balance should be struck between this principle and the basic rights of individuals to carry out activities involving the emission of a certain level of GHGs, Caney has joined a 'prioritarian' approach to the allocation of GHG emissions quotas that remain compatible with a carbon budget meeting the UNFCCC objective. Hence his formulation of another principle: 'the least advantaged have a right to emit higher GHG emissions than do the more advantaged of the world' (Caney, 2005b, p. 770). He clarifies his thinking in a later contribution: emissions should be distributed to promote the development of the world's poor and disadvantaged without triggering dangerous climate change (Caney, 2011,

p. 98). This could be done either through a direct allocation to them, or by organizing GHG emissions allowance auctioning, and allocating the funds raised to the development of those most in need.

Version (B) of the right to ownership of resources is in fact split into three different translations, by combining credible options for the source of the right and its purpose. Firstly, there is the right to use the capacity of the atmosphere to store GHG emissions (B1), which is to be seen either as a component of a subsistence right (B'1) or the subject of a customary right (B''1). Secondly, there is the right to the maintenance of essential features of the local climate and environment (temperature, rainfall, sea level); this right is supposedly rooted in the fundamental rights of local populations (B2). This raises two questions: (1) What are the features that give substance to the aforementioned claims to rights? and (2) How should a situation of conflicting rights be governed?

7.1.1 Preliminary Discussion on the Organization of Rights to Use the Atmosphere

The atmosphere is used by different types of agents for different types of use. How should user rights be allocated between users? A first possibility is to allocate rights to GHG emitters in proportion to their emissions at the time when the discussion begins. Those agents who consider themselves to be negatively affected by this distribution should then propose compensation to other emitting agents to induce them to agree to an abatement of their emissions. This is the solution supported by Chauvier (2013), in other words the principle of the 'polluted pays' for the environmental externalities created by other agents when the polluters being asked to reduce or stop their environmental externalities are only exercising a legitimate right. The second possibility is that all the inhabitants of the world are acknowledged to have a common property right to the composition of trace gases in the atmosphere. It would then be the responsibility of all producers worldwide whose processes imply emitting GHGs, but also of all consumers using fossil energy sources, to offer compensation to these inhabitants in order to obtain from them a right to a certain volume of GHG emissions. If we disregard the organizational problems and transaction costs, both formulas are conceivable. To avoid indeterminacy, a higher authority should define and distribute these ownership or usage rights.

At the level of the domestic rules implemented by a state, an indirect distribution of usage rights to the atmosphere is commonly achieved through the definition of emissions standards for various gaseous effluents, which are imposed on productive equipment and vehicles to maintain air quality at a satisfactory level. At the international level, for certain types of pollution such as those generating acid rain, regional conventions exist, including the Convention on Long-range Transboundary Air Pollution, which was negotiated in 1979 within the framework of the United Nations Economic Commission for Europe and entered into force in 1983. Its foundation is comprised of commitments by the signatory states to reduce their emissions according to a schedule, with the important counterpart that all states agree to rule out any idea of challenging the liability of other states for damage they may suffer.

Regardless of the scale, there are no cases in which all rights would be allocated either to inhabitants or to the producers or emitters of pollutants. The general solution is sought in a balance between the rights of both parties: rights to produce for one, and rights to a liveable environment for the other.

This factual statement does not suffice to settle the issue. Moral rights are distinct from legal rights, just as moral theory is distinct from legal positivism. Consequently, the narrative is the following: for moralists (Vanderheiden, Caney, Gardiner, and so on) or political philosophers (Pogge, Held, and so on), the objective interdependence between human beings in the era of economic, financial and ecological globalization is sufficient to create a moral community from which arise the rights and obligations of all. These authors believe that since the atmosphere is impacted by the GHG emissions of all humans, it is morally a common good between them. Being free and equal in rights to all other human beings, everyone has the same 'natural right' to the atmosphere and the same right to take part in deliberations to define the management regime for this resource. Being independent of legal rights, the content of which varies according to the period and the political history, this twofold natural right has preeminence everywhere and always over the contingent and imperfect norms of legal systems. Thus, it should have imposed itself on predecessors in developed countries since 1850, just as it imposes itself on contemporaries of the 2010s.

Some strong objections to this reasoning arise. It is curious to make speculative moral judgements about a pre-legal universe that would already have all the attributes of law in terms of rights and obligations, including its universal enforcement to all moral agents. Such normative discourse is aimed, in a way, at incarnating not only what legal norms should be, but also what they already are, or almost.[1] Equally strange is the postulate of the existence of a moral community organized by a complete system of rules, rights and obligations, pre-existing the formation of a political community, in an international situation in which, specifically, people do not form and do not want to form a unified political community beyond the framework of the various existing international institutions. Moreover, the distinction between moral rights and legal rights does not solve the problem of how the global climate can seriously become an object of rights.

7.1.2 Discussion of the Principles of Common Ownership and Equality Per Capita (A)

Individual equality of rights is presented by its advocates as a fundamental principle, one that is superior to any other consideration, such as the satisfaction of highly diverse specific needs on this planet. Its first postulate, with its religious belief – that God gave the Earth to humankind – which was reaffirmed in June 2015 by Pope Francis in his encyclical *Laudato si', On Care for our Common Home*, can be challenged precisely because of this origin. In fact, without deciding to what extent everyone wants to share this postulate, a decisive criticism challenges the logical link between the postulate and its derived proposition: individual equality of rights over common property.

Human life presupposes the appropriation of the material means of living and working that people find in nature. This includes the use of resources and the discharge of effluents, pollutants and waste; it is a physical principle for open systems. There are several possible ways of organizing this appropriation: maintaining common access with rules on access and exclusion; organizing a system of private rights; or the spontaneous appropriation of resources by some, provided that those who are excluded are not placed in a less favourable situation (what will be referred to later as the second Lockean Proviso) (Risse, 2012, p. 197).

Never in history have people facing a problem of regulating the use of a common resource adopted the solution of a distribution to countries in proportion to their population in the name of the presumed equal right of all human beings to common resources. For example, when marine fishing rights were being determined and allocated, an allocation of these rights in proportion to the populations of the countries or regions concerned was never envisaged. Actual solutions relied heavily on the recognition of past practices, while delimiting them, even when fishing quotas were made transferable in order to be marketable. For instance, Denmark was a pioneer in Europe, introducing the system of individual transferable quotas for pelagic fishing in January 2003, and transferable quotas for ships in January 2007 for the management of demersal fisheries (Andersen et al., 2010). The initial allocation was made free of charge, based on previous catches or previous states of fleets. This is known as 'grandfathering' in the literature.

The principle of per capita equality of rights can therefore be defended as one of the possible solutions, but it cannot claim either pragmatic evidence or a direct connection with the postulate of the common property of humankind. It does not have the status of a dominant principle and can be submitted to critical examination like any other principle.

Distributing rights and duties to populations, not states

From a cosmopolitanist perspective, the moral agents for which ideas of justice and equality are relevant are individuals, not arbitrary entities such as nation states. Arguments on equal access to common global resources should therefore result in procedures ensuring that individuals have access to their fair share. A procedure of allocation to states that do not detail how individual access would be organized would be inappropriate in terms of both duties and rights.

Equality, of course, but equality of what?

Assuming the value of equality to be universally accepted, the question arises as to what should be equalized. Scholars such as Caney (2009b; 2011, pp. 90–97) and Posner and Cass Sunstein (2009) find it astonishing to focus the search for distributive justice on an isolated good – the GHG absorption services provided by the atmosphere – independently of the access of populations to the whole set of goods on which their well-being depends. What is important for the people, they argue, is the set of property rights they have to meet their basic needs and ensure access to a decent human life. To release GHG in the atmosphere is not a fundamental human right that, in the name of equity, should be equally distributed among individuals against any other consideration such as basic needs (Hayward, 2007). These authors ask us to increase the focal length. Why not apply the requirement of equality to all rights associated with 'primary goods' or 'resources' (Dworkin, 2000)? Why not seek a regime of equal opportunity or equal capabilities?

In response to these criticisms, Gosseries (2007, pp. 282–3) believes it is justified to adopt Elster's perspective, which delimits a perimeter for 'local justice'. For Gosseries, allocating GHG emissions rights is a delimited 'local' problem that would need to be addressed as such, by disconnecting it from considerations of access to all other essential goods.

This is an interesting problem in terms of determining how we should define and delimit a question of justice, either as a local problem or as a global one, a question already considered in Chapter 2. How far can we go in extending or reducing the moral and conceptual space within which the problem is shaped? Should we try, for example, to equalize separately the distribution of emissions rights for methane and then, in turn, for nitrous oxide and then carbon dioxide? What about ocean 'carbon sinks'?

Invoking the perimeter of political climate negotiations to delimit the moral problem of climate justice, as has been suggested, does not provide a solution for all those who, like Gosseries, consider that what an individualistic ethic views as morally arbitrary elements should be dismissed, here the existence of states.

Caney's solution is as follows: it is necessary to ask to which fundamental human interests the right to emit GHGs responds. It is at this level that the value of equality regains meaning, leading the author to conclude, against supporters of allocating equal per capita GHG emissions rights, that:

> The arguments for attributing normative significance to greenhouse gas emissions entail that the latter should not be distributed according to their own distributive principle . . . There is no reason to treat the distribution of greenhouse gas emissions separately from an understanding of their moral general entitlements. Therefore the equal per capita perspective is implausible. (Caney, 2011, pp. 93–94)

Considering the overall well-being of populations, it is clear that energy needs and the derived need to emit GHGs are legitimately different from one region to another, depending on geographic and climatic constraints (think of the heating of houses and buildings), the economic situation and the actual opportunities for using non-fossil energy sources (Caney, 2009b). Equal access to a resource of one sort is not enough to ensure equity if, in reality, the possibilities of benefiting from it are very unequal. Here we find the standard debate between 'resourcist', 'capabilist' and 'welfarist' approaches.

Equality as a source of injustice?

Having previously defended the principle of equality of individual shares attributed to all people, based on a Lockean-libertarian perspective (see below), Moellendorf (2011) considered that this principle excessively penalizes the least well-off populations by adding obstacles to their development. In his considered view, these populations should be allowed to emit more GHGs than others. An unequal distribution of rights should be organized for the benefit of the poorest. Khor (2012) comes to the same conclusion with the argument that poor people and poor countries have less capabilities to make an efficient and profitable use of the emission of 1 tonne of CO_2.

The Indian economist P.R. Shukla (1999) shared the same concern about the principle of the long-term convergence of emissions rights per capita: this formal equalization could be an obstacle to the development of the developing countries in a phase of their economic growth in which they will need to emit GHGs more intensively than the developed countries. If the developing countries were obliged to buy emissions allowances on the international market, they would be unfairly penalized, warned Shukla.

7.1.3 Discussion of Conditions for the Possession of Natural Resources: Sovereignty, Grandfathering and Customary Law (B1)

Extrapolating the status of territorial natural resources

International law is based on the acknowledgement of the pre-eminent position of sovereign states that are equal in rights. Regarding natural resources, the principle of states' sovereignty over the resources in their territories has been recalled on numerous occasions, including in UNFCCC texts. As we have seen, this principle is theoretically applicable to the air column above a territory, up to the limits of outer space. It would be logical that this regime should cover all facets of the air, as for any resource. When states made use of the capacity of the atmosphere to absorb GHGs, this was strictly in keeping with their sovereign right in 1989, as was the case in 1850 and all the years in between, prior to the emerging awareness of the global climate threat created by an unlimited use.

Because of this threat, limits must be introduced in order to avoid further entering the dangerous climate zone. However, these limits cannot legitimately be elaborated on the basis that some countries have in the past taken more than their fair share of a budget that would have been defined in proportion to the population of each country. This idea and this criterion appear totally absurd here from a legal viewpoint. In other words, no notion of an 'abuse of rights to use the atmosphere' committed by states prior to 1990 is admissible.

In line with this principle of sovereignty over natural resources, the basis for the equitable distribution of GHG emissions between countries would logically be the size of the territory of each state. Such a rule would apply to emissions from the territory. This excludes a consumption-based approach, which considers the emissions associated with imports. Exclusion then has two strong motives: the emissions attached to the imported goods are foreign to the sovereign territory; and it is beyond the importing states' power to determine and control them. The factors determining a given level of emissions (the nature of the products, the nature of the technology used for production, investment priorities) are placed under the moral, political and practical responsibility of the sovereign state in which the goods are produced. This analysis leaves open the other way of challenging the behaviour of states, by mobilizing the damage inflicted on the environment of other states or areas that are not subject to state guardianship, such as the high seas or outer space.

Customary law

Some authors propose interpreting GHG emissions prior to 1990 from the viewpoint of customary law (Godard, 1992; Young and Wolf, 1992; Okereke, 2010). Their aim is to explore the normative foundation of a very frequent practical reference to the grandfathering criterion. This fact is troubling: the practical influence of grandfathering does not have an equivalent in moral theory, since attempts to defend this criterion on moral grounds are rare, as Raymond (2003, p. 166) notes. A philosopher such as Caney (2011, p. 88) sees no moral justification for this criterion: 'The only case for accepting grandfathering is a pragmatic one'. Nothing is less certain.

Luc Bovens (2011) has explored the possibility of developing a moral argument inspired by John Locke (see next section) in favour of the plausibility of grandfathering as a principle enabling the development of a more productive and efficient use of resources. More specifically, he has shown that none of the empirical differences between the atmosphere and the pastures, access to which is generally organized through grandfathering, precludes the transposition of this allocation procedure to GHG emissions. Insights from law concerning the customary emergence of a usage or ownership right are therefore very useful on this point.

To obtain a new customary right, customary law requires strict conditions: repeated and sustained use by the parties concerned without contestation from one of them; that is, a relationship expressing 'prolonged peaceful and public ownership'. The objective duration of actual use is not enough. In addition, each party must have the conviction that by making or accepting this use, or by respecting that possession, it complies with the requirements of fairness. These conditions would not be met for practices that are the subject of historical disputes, such as the occupation of certain islands claimed by two rival states or historical rivalries over the exploitation of fisheries by fleets from different countries. On the other hand, the absence of disputes over a multigenerational period gives strong plausibility to the customary value of the right of use or possession.

Significantly, groups or countries claiming recognition of the historical rights of indigenous peoples to their ancestral lands and their natural resources use the same grandfathering argument to justify the possession of these rights by these peoples. In their eyes, their claims regarding present rights to resources and practices are founded on the fact that their ancestors used the same resources, territories or practices. If such arguments are deemed admissible in these cases, it would be difficult to understand why they should be rejected without discussion in the case of access to the atmosphere.

Applying this theory to atmospheric uses, practices prior to 1990 were clearly legitimate. With the capping and distribution of GHG emissions rights in the 1990s, it was not a matter of distributing a new manna from the sky, but of regulating long-accepted practices. The adoption of a new international legal regime cannot ignore past investment in infrastructure and production tools using the technologies available at that time, which required GHG emissions; a reality about which no objections were raised internationally for a long time, almost 150 years. In fact, the demand for a transitional regime avoiding sudden breaks in GHG emissions in certain countries has come from different sides, be it the Gulf oil-producing countries, China or Brazil, and not only from the developed country camp.

However, none of the signatories to the UNFCCC can claim that these customary rights acquired by users, as observed in 1990, should be protected forever. This would amount to transforming established uses of the atmosphere into full ownership rights, in disregard of the threat of climate change. Customary rights are fragile, as they depend on recognition by all parties. Since the first international negotiations on climate change, the implicit agreement of all concerned regarding the legitimacy of unlimited uses of the atmosphere as a receptacle for GHGs has disappeared and historical customary rights must give way to another system of user rights.

John Locke and the rightful ownership of natural resources

Most debates on natural resources, and more specifically on uses of the atmosphere, give a prominent position to the contemporary interpretation of the proposals of the seventeenth-century empiricist and liberal philosopher John Locke (Bovens, 2011; Moellendorf, 2011; Risse, 2012). He indeed addressed the conditions for the legitimate appropriation of natural resources in the state of nature; that is, even before the constitution of political societies and states. Like contemporary cosmopolitanists, Locke took as his point of departure the postulate of an initial common ownership of terrestrial resources, exposed thereafter to various processes of spontaneous and unilateral appropriation by certain individuals or groups. Was this only a type of plunder that vitiated the equity of all subsequent transactions and transmissions? Was this appropriation acceptable to natural justice?

The very framework of this theory is sometimes considered inappropriate by those who, like Piguet (2014), see in the climate problem an unacceptable threat of major disturbance to the biosphere. For these authors, the biosphere must be considered as an absolutely superior good of humankind. They present it as a quasi-living organism, Gaia, and associate with it the idea of a break from life to death, an all-or-nothing alternative, far distant from a gradualism in the appreciation of the evolution of the living world. Thus, in their eyes, the first tonnes of CO_2 emitted by coal mining at the beginning of the industrial revolution were already in excess, being a source of damage to the biosphere. Piguet (2014, p. 295) states that: 'The concentration of greenhouse gases in the atmosphere has increased since the beginnings of industrialization and emissions have from the very outset exceeded the capacity of sequestration by sinks. There was no unused sequestration capacity to be used by so-called first users in 1850.'

It is the very idea of a resource known as 'atmospheric capacity to absorb GHGs' that is challenged by these authors, thus negating at the root any development on the issue of the fair distribution of this resource. Since it is not feasible for humankind to immediately cease all emissions of GHGs, it is necessary to organize a system of decreasing transitional tolerances; but this system is in no way comparable to the allocation of true moral rights. While in the 2010s annual emissions of CO_2 from fossil fuel burning are in the order of 30 Gt, Piguet (ibid., p. 253) estimates that the net sequestration capacity of the 'atmosphere–ocean–terrestrial eco-system', one that would not result in any atmospheric GHG accumulation, is around 3 $GtCO_2$, one order of magnitude less than the current figure.

Should we accept this radical view? Climatologists themselves (Allen et al., 2009; Matthews et al., 2012) call for reasoning in terms of an intertemporal carbon budget that must not be exceeded in order to stand a reasonable chance of ensuring that temperature rise in the lower atmosphere does not exceed a certain value (2°C, for example). Once we accept the idea of a carbon budget to be shared among all countries, we face a problem that brings us back to Locke's theory.

For the latter, in the state of nature, every person is both free and equal to others because no social rule or institution has yet organized significant differences between them. However, being governed by a divine order, every person is the repository of natural duties and rights. One of these duties is preserving one's own life and respecting that of others in order to preserve humankind, thereby prohibiting violence and establishing the obligation to honour promises and contracts. Rights comprehend what is required in order for individuals to fulfil their duties. They encompass the right to life, freedom, and ownership of what results from self and exchange.

It is within this intellectual context that, in *The Second Treatise of Civil Government*, Locke (1690) addresses the appropriation of natural resources, here encapsulated as land. Locke's approach is to distinguish several steps. Although originating in common ownership, land may be appropriate under four initial conditions: not being already appropriated by someone else; meeting a need for life; being limited to the actual operating capacity of the person who takes possession of it; and leaving 'enough and as good' in common for others. It was this last condition that attracted attention as the 'Lockean Proviso', a name given by Nozick. The three other conditions are nonetheless important: the occupation and exploitation criteria indicate that appropriation is legitimate if human labour merges with the resource and changes its status from a 'common good' to a 'private good'; and the criterion of need limits the extent of legitimate appropriation and is aimed at avoiding waste.

The second step is as follows: after the invention of money, Locke considered it possible to preserve value without inducing waste. The criteria of 'need' and capacity were then disregarded. This is even more true given that an owner can recruit employees. Once the appropriation process is well developed and affects significant portions of the territory, natural resources can no longer be abundant enough to ensure that there is always 'enough and as good' of them in common for those who are not yet owners. Some individuals therefore cannot access land ownership. This does not, however, make previous appropriation illegitimate in Locke's view, since the economic growth it has enabled ensures that those who do not own land benefit from living conditions that are at least as good as if the land had remained the common property of all. This last formulation, which I call the Lockean Proviso 2, is the distant ancestor of Rawls's 'principle of difference': for the latter, once basic liberties are ensured for all, inequalities in the possession of primary goods are fair if they make it possible to improve the lot of the least well-off.

Thus, for Locke, natural resources can be legitimately appropriated via production under two conditions: either 'enough and as good' remains for the others (Proviso 1); or the development of economic activity made possible by appropriation ensures that those who are deprived of access to these resources are not disadvantaged by the loss of access to the commons induced by the scarcity of land and the appropriation process (Proviso 2).

An adaptation of this type was required to ensure the feasibility and legitimacy of a society giving a prominent position to private property. It is therefore necessary, from Locke's point of view, to interpret the Proviso 1 as being relevant only at the beginning of the appropriation process. It applies only to the contemporaries of these beginnings, such as when the first settlers arrived in America to appropriate land to cultivate; putting aside here the embarrassing fact that this land may have been the territory of indigenous populations. For the rest of history, it is within the logic of the appropriation process that resources that have still not been appropriated become scarce.

To give a transhistorical meaning to the Proviso 1, to extend its benefits to all generations to come constitutes a senseless deviation that would be incompatible with Locke's thinking and with the foundations of modern society, with the prominent position given to property and the market economy. We are confronted with the following alternative: either this

Proviso could be respected and there would be no place for the emergence of property rights and markets – for what purpose, since unlimited abundance would reign? – or it cannot be respected over time, and a market economy can then take off to stimulate the productive use of resources that are increasingly scarce. Thus, within a Lockean framework, the intertemporal justice of the use of natural resources cannot obey the same criteria as synchronic justice in the beginnings. This point was missed by all commentators of Locke who took part in the discussion on the ethics of climate change. For instance, the claim by Neumayer (2000) for equal opportunity ensured for each person, regardless of space or time, is futile, in that it negates the deep sense of history: successive generations do not live in the same world.

With these ideas, let us see how Locke's theory is used by some contemporary scholars to address the climate change issue. To clarify the grandfathering criterion, Luc Bovens (2011) found in Locke a source of validation for the conditions of past appropriation of the atmosphere through GHG emissions, but also established two limits: (1) appropriation conditions must not undermine the means necessary for the basic subsistence of any person; and (2) the legitimacy of appropriation collapses in the event of wastage of the appropriated resource. In other words, Locke's thinking on the state of nature would give priority to minimal access to the atmospheric resource in order to cover the basic needs of every human being, and subject its appropriation to the constraint of efficiency in its use.

Another author to re-examine Locke's theory, Moellendorf (2011, pp. 106–9), takes up the postulate of the original common ownership of natural resources, but interprets it without further precaution as giving every person equal rights to the atmosphere. However, historically, this appropriation has led some to consume far more than their 'equal share', in apparent violation of the Proviso 1. As for the Proviso 2, Moellendorf expresses scepticism that it was respected at all. For him, there is no evidence that the economic growth enabled by excessive CO_2 emissions since 1850 has generated such benefits for the victims of climate change that they are better off than if this economic growth had never happened, and the equality of use of the common resources had been preserved. He concludes at the level of principles that Locke's theory reinforces the case of those who demand compensation from countries that have emitted more GHGs than their fair share.

Is Moellendorf's argument convincing? As already discussed concerning the 'non-identity problem', it is well established only in appearance, and comes up against several objections. It can, firstly, be attacked for its disregard for the temporal and historical dimensions of the problem, since it postulates the possibility of defining the equal share of every human being in terms of access to the atmosphere for an unlimited historical period. Yet the level of this equal share was not known to Europeans in 1850 or 1930, nor was it known to the Chinese or Indians in 2015. This is due to uncertainty about demographic change in the very long term. In 1930, when the world's population stood at 2 billion people, no one knew that there would be more than 7.3 billion people in 2015.

Secondly, the validity of Moellendorf's conclusion depends on the comparison with a counterfactual scenario concerning the situation of the victims of climate change. Such a scenario is problematic. What would be the state of the human population (lifestyles, level of development) in a scenario in which humanity had from the outset preserved equal access of all people to the atmosphere? Different speculations can be put forward, but they all encounter the following problem: if humanity had chosen to maintain equal limited use of the atmosphere by individuals under a common ownership regime and renounce economic growth rather than accepting an unequal appropriation of the atmosphere for the immediate benefit of some of its members through economic growth, the present human population would not be composed of the same people, either in numbers or in 'quality'. In other words, the alleged contemporary victims of climate change, who Moellendorf doubts have benefited in any way from the economic growth of the past two centuries, would not exist if this historical process of industrial development had not occurred. This is because it has had a profound influence on all aspects of human life across the Earth, including hygiene and medical care in less-developed countries.

It is therefore impossible to compare the situation of these people as the victims of climate hazards with what their situation would have been in a counterfactual hypothesis of common ownership of natural resources maintained throughout human history, and specifically over the last two centuries. In this intertemporal exercise, it is not possible to use the test of the second Lockean Proviso due to logical impossibility. Locke is also caught in the non-identity trap. His provisos can only apply to a given population with a fixed identity, which limits the temporal scale of their validity to short periods during which the alternatives envisaged could have no effect on the conditions for the procreation of new human beings. This confirms that it is indefensible to apply the Lockean Provisos over a historical period.

In any event, one thing is certain: were it valid, the Lockean Proviso 1 would have been strictly respected until 1990. The use of the atmosphere through the emissions of GHGs had then left 'enough and as good' for others, since no limit had yet been set on these uses. Moreover, prior to that date, no country had experienced any significant deterioration in its situation due to the excessive global use of the atmospheric capacity to absorb GHGs. This is a historical dividing line that no one should forget when drawing up the future regime of rights to use the atmosphere.

7.1.4 Discussion of the Idea of the Right to an Unchanged Local Environment (B2)

Intrinsic fragility

The thesis of the existence of the natural rights of populations to the key characteristics of their local climate and physical environment is fragile, if not wholly dubious. Historically, no equal right to the local climate has ever been defined or acknowledged, either at the nation state level or on a global scale. There are good reasons for this, in addition to the absence of a global state. The fragility is at first intrinsic to the idea of natural rights, which are deemed universal and independent of any formal recognition by societies and their laws. This idea is inconsistent with the principle of state sovereignty and with the democratic idea, which both assume that the definition of rights is a matter for the instituting power attached to sovereignty. More specifically and more central to the present discussion, the idea of natural rights to a given local environment expresses a fundamental misunderstanding of what is meant by the concept of a natural environment.

A natural environment, including climatic conditions, cannot be understood as a list of goods and conditions that people can control as they wish. Environmental characteristics can hardly be encapsulated in stable, intangible properties that can be subject to rights: wind, rain, storms, temperature, coastal erosion by the sea, permafrost thawing, biological vectors of disease, and so on are largely beyond local control. The paradigm of interaction is quite different from that of control. Yet the claim of a 'right over' or a 'right to' something at least requires that 'something' to be under the control of human action. It is important to insist on the gap between the features of natural environments and the qualities required for something to be the object of a human right. This issue of control is enlightened by systems theory.

The concept of the environment for self-organizing, open systems

Self-organizing, open systems have an environment with which they interact in an essential way for their survival and development. They find in this environment vital resources, information, a receptacle for waste, but also disturbances and threats to their development and survival. When we think of the fundamental features of the environment of such systems, we find (Godard, 1995; 2015, Chapter 2) that:

1. The environment category connotes an idea of exteriority, in the sense that it marks both the limit of the domain of practical and informational control that these systems have acquired, and the gap between their domain of action, which extends into their environment, and their much more restricted domain of self-reproduction (all interrelated processes and functions actively maintained over time in order to preserve their integrity). Human rights belong to this restricted domain of self-reproduction of human societies and depend on a capacity of control. The relationship with the environment cannot be kept under such complete control.
2. To the extent that the autonomy of these systems is based on their capacity for self-reproduction, and that self-reproduction depends on the maintenance over time of a complex set of linkages and loops between components and processes, their environment is then made up of beings and forms of organization (living organisms, ecosystems, and so on) whose reproduction is not actively and regularly incorporated as a functional component into the organization of the systems themselves. The active movement of self-production and reproduction of these systems is also the active mechanism for reproducing what may be called the 'externalization of the environment': the process by which systems reject the systematic reproduction of a component or a process that will therefore be part of their environment. The autonomy of self-organizing systems and the externalization of their environment are indeed two sides of the same coin.
3. Considered at meta-levels at which coherent organizational rationales manifest themselves, the environment is also an autonomous pole that creates its own meaning, thus representing a permanent source of uncertainties and surprises for the systems that interact with it.

Although self-organizing systems cognitively and physically construct their environment through what they are themselves, these environments also represent for systems a source of disturbances and surprises that they do not control, and to which they must adapt. Some of these disturbances are a feedback loop from the process of externalization that systems imposed on their environments. Thus, the reproduction of the environment can never be taken for granted. In fact, the primary process is the one by which the actions of systems undermine the conditions for the reproduction of their environment. When finally secured, this reproduction appears as the risky and provisional outcome of externalization caught up in the game of adaptation and evolution. The gap between action and reproduction places the relationship between systems and their environment in an open history that can generate novelty, and not only repetition.

It opens the prospect of a coevolutionary trajectory between systems and their environment, but this can fail.

All this corresponds well to the key features of the relationship between human society and climatic conditions. The main variables of the description of climatic data cannot enter the category of a factor controlled by human societies, whether local or international. With climate events, humankind faces fluctuations, instability and uncertainty that are only tempered by some constraints due to geophysics (the changing seasons, and so on). Looking back over long periods, scientists can identify regularities in outcomes, enabling the characterization of different types of regional climates, and processes such as El Niño in Pacific waters. But nobody can control climatic events; even forecasting is difficult. Regarding the future evolution of climate change, scientists can only model the influence of GHG emissions and atmospheric concentration on general statistical trends on a regional or global scale. Climate dynamics combine natural factors of change and human actions with multiple negative and positive feedbacks. The idea that local climate characteristics could be stabilized forever everywhere is a myth that would drive environmental protection policies in a radically wrong direction. Within such an objective context, the idea supported by moralists of a universal right to local climate stability constitutes a conceptual and scientific monstrosity.

A detour via substantive law

To go further and make the point clearer, a supplementary solution is to look at substantive law. Climate moralists, such as Shue and Gardiner, have often solicited certain norms of substantive law to support their proposals, although they underline the fact that moral thinking and law are two different universes. For example, whether rightly or wrongly, they have looked for support for their proposals on historical liability in the 'strict liability', 'no-fault liability' and 'polluter pays' principles.

What does substantive law tell us? The International Covenant on Economic, Social and Cultural Rights, which was adopted by the United Nations General Assembly in 1966 and entered into force in 1976, but was not ratified by the United States, makes no reference to rights relating to the environment. It does not provide for procedures whereby people who feel that some of their rights have been infringed, such as rights to food, housing or access to care, may have such violations recognized by an international body.

Regarding the fundamental rights invoked in the climate debate (life, health, subsistence), their meaning and scope must not be misunderstood: the right to life cannot be interpreted as a right not to die or as a right not to be exposed to the hazards of life. The same goes for the right to health, which cannot be understood as a right not to be exposed to various threats (viruses, bacteria, accidents, and so on). These rights are known as rights of claims: 'rights to'. They generate a general obligation for the public institutions in the country in which people live to ensure the protection of those citizens' lives and public health, with specific consideration for the least well-off, or the neediest, because these population categories are often the most exposed to all sorts of threats and are the most vulnerable.

These rights generate for states obligations of means, not obligations of results. They are to be applied through the interpretation that can be given to them in the context of the public action of each state. General constraints on the deployment of this action must be considered, particularly for budgetary constraints and those linked to available human resources. In their accomplishment, these rights cannot but be in phase with the state of the society in which they are sought to be acknowledged. They are not the same as an exclusive right of ownership, which is opposable to third parties since it concerns a fundamental liberty. A complaint resulting from a breach of a right of claim is addressed to the state in charge of its protection, whereas a complaint for harm in case of violation of a 'right-freedom' is addressed to the author of the damage. If something similar to a right to an unchanged climate were acknowledged, it would be the responsibility of the public institutions of the country in which the people live to provide a response which, by necessity, would have to take account of other pressing demands on public means. In this regard, this could add a supplementary element to the issue of international distributive justice, in case the public means of the affected state cannot address this additional burden.

A fragile link to essential human rights

The 'right not to suffer the disadvantages created by global climate change' advanced by Caney has no equivalent in substantive law, but this is not a sufficient objection for those who pursue a moral reflection. Certain declarative texts with symbolic significance have a certain kinship, such as Article 1 of the Charter of the Environment included in the French Constitution since March 2005, which states that: 'Everyone has the right to live in an environment that is balanced and respectful of health' (Bourg, 2006, p. 236). The right defended by Caney, however, is much more precise and demanding. It implies the recognition of a right to a climate that is not impacted upon by climate change, the previous state of the climate before 1990 being used as a reference to qualify the changes in terms of damage to be compensated.

The nature of the link invoked between fundamental rights and the right to an unchanged climate is clearly not established. Nevertheless, many analysts and activists (Mary Robinson Foundation, 2011; Shue, 2011;

Caney, 2011; Koivurova et al., 2013; Piguet, 2014) affirm the unquestionable existence of a strong link between human rights and environmental protection. This point needs to be further addressed. The lawyer Philippe Cullet (1995, p. 26) wrote: 'On a factual level, it has already become apparent that preservation, conservation and restoration of the environment are a necessary and integral part of the enjoyment of, inter alia, the rights to health, to food and to life, including a decent quality of life.' However, the author (ibid., p. 31) recalled that:

> The holders of a right to environment can neither claim a given state of the environment, nor a perfect environment (this has never existed since humans appeared on earth), nor a local environment similar to other places in the world, as the unequal distribution of resources does not allow for the existence of equal environmental conditions everywhere.

Between a given state of the environment affected by climate change and fundamental human rights, there is a distance that reduces the strength of the link. None of the three fundamental rights mentioned (life, health, subsistence) depend directly and solely on the climate: they depend first and foremost on the level of development of the productive means in the country; on the economic and social rules in force; on public health policies; and on social security and solidarity institutions working for the most disadvantaged people and the victims of natural disasters.

For example, the impact of heatwaves on the mortality of the elderly depends, of course, on these climatic events, but even more so on social functioning (whether or not these people are socially isolated, their level of equipment at home, the availability of places to receive them in specialized establishments during these episodes, and so on) and public policies for prevention and precaution. We need to contextualize evaluative judgements about climate events and the changes induced in local environments. We should ask ourselves, for example, whether farmers have access to reasonably cost-effective insurance to cover losses due to climatic hazards, whether there are accompanying policies to help relocate populations whose habitats are threatened by rising water levels, whether security policies have been put in place to improve local resilience to climate events, and so on.

The final impact of climate change on people and local populations is mediated by the fabric of the economic, social and political processes in each country in terms of insurance, solidarity and security. It is the prerogative of sovereign states to organize this for their people. The logic of fundamental rights of claims extends the obligation to respond to the changing environmental conditions through various policies and services, but does not directly trigger claims for redress at the international level.

Climate change affects the circumstances in which their rights are implemented but cannot, as such, be sufficient to characterize an infringement of those rights.

On the other hand, states in which populations exposed to climate change may reside may consider that the additional expenditure and attention required of them because of this new environmental condition represents for them not only an additional burden, but harm attributable to all those who will be held responsible for climate change. In this regard, many states, apart from the poorest, will find themselves jointly responsible for assisting or compensating for the financial unbalances indirectly generated by climate change in the public finances of the countries that are the most exposed or the most vulnerable to climate events, or the poorest in terms of the public means available. Their responsibility would be differentiated, as reported in Chapter 1.

Some authors, including Caney (2005b), Shue (2011) and Moellendorf (2014), propose organizing the distribution of emissions quotas to countries in inverse proportion to their wealth or in proportion to the priority needs of the most deprived. The purpose is to ensure that they have not only a minimum level of subsistence, but a decent life and development without additional obstacles. In short, emissions quotas are used as a resource for development for those who need it most. This is generous and demonstrates solidarity with the least well-off worldwide, but remains artificial in its link with the climate issue. Why should we want to organize the promotion of development through the climate protection regime? Is there no more direct channel for doing so? The advantage of these proposals is that they draw attention to the risk that the most vulnerable populations may see their prospects for development stifled either by climate change or by policies to combat this change. Safeguards must be implemented. Beyond that, moralists, who are so eager to purify their reasoning of any element deemed to be contingent and arbitrary, should find it inappropriate to manipulate or divert international action against the threat of climate change from its main purpose.[2]

In retrospect, one can only be struck by the recurrence of the dual temptation to naturalize the physical, health and economic situation of local populations, and to humanize physical phenomena (a tornado, a heatwave, a landslide) and environmental changes (cliff erosion, the contraction of glaciers, and so on). The idea of climate harm is presented as the outcome of a ballistics type of phenomenon, implying a direct linkage between specified human actions (GHG emissions) and a specified harm, whereas eventual harm is fundamentally marked by local and national economic and social conditions and by the policies of each state. Symmetrically, many seek a way to impute an exclusive human responsibility for the variability and dynamics of natural events.

How to arbitrate between legitimate conflicting rights?

To remain within an approach in terms of rights, the situation created by the threat of climate change brings out two contradictory types of rights, which are opposed even when they are the prerogative of the same person: on the one hand, we find a claim to unchanged local environmental conditions (type B2); and on the other hand, there is the right to use the resources of the local environment, including the right to emit GHGs (types A and B1). Even if an overall emissions ceiling is set in line with the 2°C objective, this will not prevent the occurrence of certain climatic events that may affect or jeopardize the supposed rights to an unchanged local environment. The idea put forward by authors such as Shue, Caney and Moellendorf is that the violation of type B2 rights should be compensated by applying what they call the 'polluter pays' principle.[3]

Fully allocating the costs and moral discredit to GHG emitters would be tantamount to recognizing the exclusivity or superiority of the rights of those claiming an unchanged climate over the rights of emitters to use the atmosphere, while the latter have emitted and will continue to emit in accordance with their rights. Since Coase's (1960) theorem, we know that the situation of the rights holders is symmetrical in relation to the environment: the terms 'polluters' or 'victims of pollution' can only be used once the respective rights of all have been defined. Before that decisive step, there are just conflicting uses. Thus, a problem of this kind calls for a fair and balanced treatment of the contradictory interests involved. As already noted, Chauvier (2013, p. 109) proposed adopting the 'polluted pays' principle for conflicts between legitimate rights:

> The only solution to harmonizing the rights of victims and those of agents is to grant victims of external damage not a right of veto, but a right to prohibit the other agents from harming them, subject to the condition of providing compensation to those agents for the cost of their abstention, or of bearing the cost of the external immunization of the actions of these agents.

We may also consider the principle of reciprocal compensation, put forward by economists sensitive to the incentive dimension of rules (Siebert, 1987, pp. 154–5): the victims of damage bear the costs incurred by emitters in reducing their emissions, while the latter compensate the victims for the damage suffered and the obstacles to be overcome. In any case, these compensation solutions would only be implemented in parallel to emissions responding to an allocation of rights to a defined global carbon budget deemed acceptable for a given period, and not for emissions beyond allocated rights.

7.2 EQUITY

In its chapter on equity, the 1995 Intergovernmental Panel on Climate Change (IPCC) report (Bruce et al., 1996) identified several broad categories of issues: (1) international equity regarding the impacts of climate change; (2) international equity regarding the burden to limit climate change through emissions abatement or sink enhancement; (3) domestic equity within countries regarding adaptation to climate impacts as well as burden-sharing for mitigation policies; (4) equity in international negotiation processes and governance; and (5) intergenerational equity. I have already considered in previous sections issues (1) and (5) and touched upon issue (4). I will put aside issue (3), which is not the focus of this book. This leaves issue (2) to be examined in what follows.

The first principle laid down in the UNFCCC for allocating rights and efforts to reduce emissions is equity, as different from criteria of responsibilities and capabilities. The text does not give any additional indication as to how this principle should be interpreted. There are two possible ways of doing so. The first is aimed at finding an objective and impartial criterion for sharing a common resource among equal parties, without considering the various sources of differences between them. There are two candidates, depending on views of the meaning taken by the allocation at stake: is it a matter of allocating rights to people, or of distributing burdens between them? In practical terms, the two views are equivalent – burdens amount to the difference between actual emissions and rights – but each one points to a different criterion. If it is a matter of rights, the attractor mostly favoured is an allocation based on population. If it is a matter of burdens and efforts to be shared, the attractor is an equal percentage of gross domestic product (GDP).

With the second interpretation, equity implies considering the relevant specificities of each party in terms of the extent of unmet needs, income distribution, wealth, and past efforts to control GHG emissions. Several sub-debates arise. One concerns the scope given to the issue of redistribution. Should the new international climate protection regime be used as a means of redistributing global wealth, or should this regime simply take account of existing inequalities, while retaining its central climate goal? Another debate concerns the main target of redistribution, if any. Not everyone considers that the relevant concern when allocating emissions rights should be reducing inequalities. Thus, on one side, we have those who above all want to combat inequalities as such; and on the other side, those who want to improve the situation of the people most in need. For the latter (Copp, 2005; Moellendorf, 2014; Shue, 2014), the concept of basic needs should be clearly prioritized and a distinction drawn between subsistence and luxury issues. For instance, being distrustful of trading mechanisms, Shue proposes the share of emissions allowances allocated to countries and destined to cover the basic needs of people not to be allowed to enter an emissions trading scheme.

Being primarily concerned with the needs of the most disadvantaged should imply inventing procedures to focus allocation of allowances and means on specific groups within each country. This is problematic for an international coordinating body, since it implies interfering in the domestic affairs of countries. Here, we again find an opposition between those with a cosmopolitanist outlook and those who follow a less intrusive concept of international justice. Representative of the latter, Miller (2007) attached the greatest importance to the principle of national self-determination, which makes states the relevant agents of international distribution. In his view, international wealth inequalities are not unfair if they do not compromise the satisfaction of basic needs, are not the legacy of past international injustices and do not significantly affect the capacity for political self-determination of the poorest countries. The redistributive dimension is not specifically eliminated, but it is no longer theorized in terms of combating international inequalities as such.

To take the redistribution of wealth as a primary criterion for the allocation of emissions quotas would radically change the motivation of the policy. As Carsten Helm and Udo Simonis (2001, p. 7) pointed out in another context: 'Our motivation is thus no longer the equitable distribution of a common resource . . . but the realization of an income transfer . . . even though the former would be justified without the latter'. Adopting the framework of a 'local justice' issue, these authors therefore proposed, in the name of equity, a combination of an equal allowance per capita for the initial allocation phase, and rules for the subsequent international exchange of quotas to satisfy one or several of the following criteria: Paretianism aimed at minimizing collective costs; the absence of envy (each person prefers their basket of goods to that of others);[4] meeting the individual interests of all (no agent should feel themself to be a loser in relation to the *ex ante* situation); and no agent should receive more benefits than they would have obtained had they been alone in using the common resource (upper limit set by the stand-alone criterion)[5] (ibid., pp. 13–14; Helm, 1999). Regarding mitigation burdens, with these criteria, no country should take an additional benefit due to the others having to bear their climate burden.

By combining individual interest and stand-alone criteria, Helm (1999) suggested a practical way to proceed: as long as GHG emissions in the developing countries are less than their 'fair shares' – here identified with per capita equal allocation – the developed countries would have to pay the full incremental cost of emissions reductions in the developing countries, but not more. Thereafter, when their emissions reach their fair shares, the developing countries can still sell emissions quotas to the developed countries up to the point of equal value across the marginal abatement costs known by all countries. But they should not keep the extra value obtained beyond the stand-alone criterion. Naturally the outcome of the procedure described depends on what those 'fair shares' are. This issue remains unresolved from the viewpoint of justice.

The procedure just described is similar to the one that several experts have proposed to incite the developing countries to join an enlarged Kyoto Protocol (Philibert and Pershing, 2001; Guesnerie, 2003). The Kyoto Protocol was based on the developing countries having no quantitative restrictions on their emissions. How should we interpret this international structure of commitments in relation to equity? There are two possible arguments. Firstly, due to unmet basic needs in the developing countries, equity implies that the latter should be relieved of all responsibility and solidarity regarding their GHG emissions. Secondly, since their level of GHG emissions is well below their fair share of rights, equity simply requires the other countries to allow them to fully enjoy their rights. Putting aside any economic criticism of the evident inefficiency of such a state of rules, we can consider both justifications as being fragile or dubious.

Where the first is concerned, we can admit that economic conditions should legitimately count in modulating the definition of duties and obligations; but they cannot completely exempt certain countries from any responsibility for their actions at a time when most countries are attempting to elaborate a common regime in which each country assumes its share of the new burden. Duties may be differentiated in the name of equity, but there is no justification for some countries having no duties at all, unless they cannot stand as sovereign states. As for the second justification, the presumed overabundant rights to emit GHGs, which would have explained why the developing countries made no commitment to mitigation actions, are absent from the UNFCCC. The latter only recognizes their right to sustainable development; in other words, a type of development taking due account of environmental protection goals, including the climate issue. It follows that, as far as international equity is concerned, the structure of the Kyoto Protocol is so unbalanced that its unfairness cannot be hidden.

To validate the Kyoto structure despite everything, it would have been necessary to put aside or disregard the demands of equity, strictly speaking, for the benefit of the other two criteria ('common but differentiated responsibility' and 'capabilities') and to provide a strong foundation for the latter. Regarding the differentiated responsibility criterion, it failed in placing all responsibility on the shoulders of the developed countries. As seen in Chapter 6 on historical responsibility, the debate between scholars has seen two camps emerge to interpret the historical responsibility of states. For the first camp, which has an extensive conception of responsibility, based only on causal links and disregarding any consideration of the subjective conditions of the presumed agents (identity, intention, knowledge of consequences, margin of choice available, etc.), the mere observation of cumulative emissions over decades is enough to characterize a moral responsibility. For the second camp, the subjective conditions of the moral agent are important and raise serious obstacles detailed in Chapter 6.

In 2016, members of the first camp cannot ignore the reality of shared causal responsibility since cumulative GHG emissions show approximate parity between the developed and the developing countries. As for the members of the latter, having argued that the conditions for liability to be challenged were not met prior to 1990, they should admit the full responsibility of all countries as of that date, with the developing countries' share having become preponderant at 58 per cent by 2013 and having continued to grow since then.

The only remaining justification for a Kyoto-type structure that exonerates the developing and emerging countries from obligations to be concerned about their emissions levels is the criterion of capabilities. This may be valuable for some developing countries stricken by hostile environmental conditions, poverty and war, but certainly not for the large countries on the three continents of Africa, America and Asia (Brazil, India, China, Nigeria, South Africa, Saudi Arabia, South Korea, and so on). Almost 20 years after the adoption of the Kyoto Protocol, the Paris Agreement has marked a clear improvement from the equity viewpoint through its universal approach requiring all countries to present their plans for mitigating their emissions.

7.3 THE ABILITY TO PAY

The 'ability to pay' offers a principle that is often given a simple translation: individuals or states should be asked to contribute to the burden of GHG emissions reductions in proportion to their income or to an indicator of wealth. This approach, inspired by the 'deep pocket' doctrine, which is well known in the field of insurance, lacks any specific justification in relation to the climate issue. Of course, a very unequal distribution of income within and between countries leads to a very unequal distribution of emissions: people emit more GHGs in absolute terms if they have a high income. There could be nearly two orders of magnitude between the top 10 per cent and the bottom 10 per cent worldwide.[6] However, two supplementary ideas should be underlined. For the same level of wealth per capita, emissions may vary greatly, by a ratio of 1:2, as shown by North Americans and Europeans, or even 1:3 at the scale of countries, and much more at the scale of individuals. Moreover, if we consider the intensity of emissions per unit of income, we see a general trend of a slight decoupling of income from emissions, which entails a decrease in emissions intensity for higher levels of wealth: the wealthier people are, the less they emit per dollar or per euro. On this point, Lucas Chancel and Thomas Piketty (2015, p. 21) note that: 'Elasticity values for consumption expenditures to energy and CO_{2eq} collected by Chakravarty et al. (2009) from 17 countries and time periods, range from 0.4 to 1 for energy and from 0.6 to 1 for CO_{2eq}, with most results in the 0.8–1 range'. For their own work, these authors take an income elasticity of GHG emissions of 0.9. This means that, with the same structure of behaviours and supply, a redistribution of income to the least well-off will increase the overall level of GHG emissions; a paradoxical result that should be stressed since some parties may be tempted to assume that reducing inequalities through redistribution of wealth would also reduce emissions levels.

In any case, there is something confusing in these presentations. GHG emissions were not measured at the level of the consumption practices of people representative of different income groups in different countries, but were: (1) assumed to obey a functional relationship to income; with (2) all emissions from production, imports and government activities being attributed to consumers' lifestyles. This has consequences. For instance, all carbon inefficiencies of production in countries that export goods are imputed to consumers in importing countries. The authors acknowledge that indirect emissions are more important and more correlated to income than consumers' direct emissions. This means that the alleged distribution of emissions does not reflect the real localized emissions that can be imputed to different categories of agents in different countries, and that authors have been drawing an apparent landscape of emissions distribution that was in fact a quasi-landscape of income distribution.

If we consider the issue from the viewpoint of international climate justice, these data are not very useful. They are presented within a cosmopolitanist framework based on income groups worldwide. This would be valuable if international civil society wanted to control and transform the domestic structure of income distribution in each country. But such aggregated data on per capita income and emissions do not sufficiently consider the moral and political significance of the organization of humankind into around 200 sovereign states. One reason justifying the existence of those states is that it enables the emergence of a plurality of self-determined political communities that elaborate their own conceptions of the just and define their own equilibria on the classical trade-offs between maximizing output and the fairness of its distribution. The message put forward by a cosmopolitanist approach is that we should make the rich pay, not only at the country level, but within each country. The proposal thus presents itself as direct interference in the domestic affairs of each country. For international justice, directly relevant inequalities are those between countries.

If the criterion of ability to pay is intended to catch the responsibility of countries in view of international burden-sharing, it would be simpler to consider the total emissions level of each country rather than income. If this criterion is intended to reflect the capacity of countries to transform their productive basis towards a low-carbon economy, the alignment of the capacity criterion with the level of GDP proposed by several authors, including Moellendorf, is too crude to be convincing. It does not consider the irreversibility or rigidities characterizing the geography and modes of development adopted by a country, whatever its level of wealth. Nor does it consider the existence and magnitude of the economic surplus (what remains of income once the current expenditures necessary for ensuring the current production and consumption of the population are covered) enabled by economic growth. It is the size of this surplus more than the absolute level of GDP that reflects a country's ability to redeploy its productive organization and its habitat through investment. In other words, a country deemed to be rich, but which has zero or negative growth, has poor means to reconfigure its productive capital and infrastructures, which form the backdrop of its GHG emissions. Symmetrically, a country considered to be relatively poor, but which has a high rate of economic growth, has considerable margins for guiding the formation of its physical capital. It is striking that many of the authors who consider ways and means to apply the criterion of equity reason from the binary opposition between the rich and the poor, which leads them to ignore the importance of economic mediations underlying moral and political choices, and the subtleties of the problems raised by climate change: there will be many indirect victims of climate change in the developed countries; most countries, and especially emerging countries like China, will be both major polluters and major victims. Overall, the ability to pay would make more sense in terms of organizing the duties of international assistance to the most vulnerable to climate change than a requirement of equity to allocate burdens to reduce GHG emissions.

7.4 THE PRINCIPLE OF RESPONSIBILITY

Here, the interpretation of the principle of responsibility is straightforward: future emissions reduction efforts to remain within a given global carbon budget should be allocated in proportion to the countries' responsibility in creating the problem. However, as already seen, there are several concepts of responsibility (causal, outcome, moral, remedial) with different conditions. The year 1990 was chosen as the main reference for setting the targets assigned to countries by the UNFCCC and the Kyoto Protocol. That same year also saw the publication of the first IPCC report. The question of accountability thus emerges in different terms both before and after 1990. From 1990 onwards, no one could use the argument of ignorance about the existence of a major climate threat to avoid responsibility.

The scope of historical responsibility prior to 1990 was considered in Chapters 5 and 6 to test the consistency and validity of the concept and the associated norms of justice. Two aspects were considered, one in relation to the prospects of damage due to climate change, and the other regarding the claim that past emissions exceeded the countries' fair shares. The second aspect deserves additional comments and is of concern in this section.

7.4.1 Repayment for Past Emissions Exceeding Fair Shares

The question here is whether or not states that have emitted over time more GHGs than others should receive a smaller share of a carbon cap in the future. According to the definition of the concept of 'ecological debt' given by Erik Paredis et al. (2009), the element constitutive of a historical debt to be repaid in the form of lower carbon emissions rights in the future is 'the exploitation or use of ecosystems and ecosystem goods and services over time by country A at the expense of the equitable rights to these ecosystems and ecosystem goods and services by other countries or individuals'. In this vein, Neumayer (2000) advocated equal per capita emissions combined with historical accountability to take account of historical inequalities in per capita emissions. This means that the more a country has emitted in the past, the fewer rights it will receive in the future. Similarly, Tejal Kanitkar et al. (2009, 2010) developed a rather straightforward concept of carbon space and carbon budget. According to them: 'Since the available carbon space is part of the global atmospheric commons it is also evident that every nation's fair share of carbon space is proportional to its share of the global population' (Kanitkar et al., 2010, p. 3).

Typically, their approach follows different steps: (1) determining a total carbon space budget over a reference period – they chose the 2000–2050 period; (2) computing the actual cumulative budget of emissions for each country since past dates (1850); (3) determining the fair historical carbon budget share of countries on an equal per capita basis; (4) determining emissions cuts to be achieved based on whether a state is above or below its fair share – the larger the gap between the actual budget share that has been consumed, the more stringent the cuts to be made in the coming period; and (5) allowing growth for countries that are below their fair share until they reach their fair share. Khor (2012) held the same position. Thus, states with cumulative emissions that are above their fair share of the intertemporal carbon budget should arrive at negative emissions: that is, they should dramatically reduce their current emissions as well as finance emissions cuts elsewhere.

According to a cosmopolitanist priority perspective, Meyer (2004, 2013) suggests a different way of giving room to historical responsibility in allocating fair access to the carbon space. As previously stated, Meyer places his proposal within the framework of a general distributive requirement, which is indifferent to issues of the wrongness or injustice of past emissions. Historical emissions therefore count for him in only two ways. Firstly, he assumes that a principle of equal per capita benefits from emissions rights should be understood not for each point in time, but in a cumulative way for the whole lifetime of people. Presently living persons should thus receive emissions rights for their future activities in a proportion that would take into account the amount of emissions from which they have already benefited since their birth. Because the living inhabitants of the developed countries have on average already emitted far more than those of the developing countries, they should receive less than an equal share for their remaining lifetime. Secondly, the currently living inhabitants of the developed countries are still enjoying some lasting benefits from past activities that had GHG emissions as by-products. Thus, the past emissions rights of predecessors are still providing benefits to currently living people. According to a norm of universal equal per capita benefits gained from emissions rights, those who presently enjoy greater benefits from past emissions should thus be granted proportionately fewer new emissions rights in the future than the strict equal share.

Both the reparative and the distributive foundations of these rules for carbon budget allocation have been disputed in several ways, with varying degrees of importance. The historical component is exposed to various arguments presented in Chapter 6. I will not repeat them here. But other arguments have been exchanged, which should be mentioned.

7.4.2 What about Terrestrial Sinks?

As put forward by Megan Blomfield (2013), the capacity of the global ecosystem to absorb GHG emissions is not limited to the atmosphere. We should also consider ocean and terrestrial ecosystems that work as carbon sinks. These sinks absorb roughly 50 per cent of gross GHG emissions. The interesting point is that terrestrial sinks, which absorb some 25 per cent of total gross emissions, are not global common resources. They are placed under the direct control of sovereign states, like other natural resources. Due to states' unequal sink resources, equal per capita emissions rights would clearly be unfair in that they would ignore the unequal net contributions of states to emissions control through their sinks.

7.4.3 A New Case of Retrospective Illusion

From an intergenerational viewpoint, the norm of equal benefits from lifetime emissions rights (Meyer, 2013) suffers from retrospective illusion because of the uncertainty regarding the future evolution of the population. A fixed carbon budget and the norm of equal share per capita cannot be established before the end of the historical period under consideration, when population data is fixed, which makes it impossible to postulate that previous generations should have known and acted accordingly. Uncertainty is even greater if the time under consideration is extended to several decades and, in some statements, forever. The 'fair share' accruing to each generation depends on future decisions by future generations to have children or not, and on future living conditions that would affect their life expectancy. For a timescale of one century, world population levels at the end of the period can vary by a ratio of 1:3, with the same proportion accruing to the equal sharing of emissions rights. This is not a type of uncertainty that could be removed with the advancement of knowledge.

7.4.4 Responsibility after 1990

The question remains as to the responsibility of populations and states after 1990, as a basis for determining the future emissions rights of countries. From a quantitative point of view, the stakes are not insignificant, since cumulative CO_2 emissions over the 1990–2012 period are of equal weight to those in the 1850–1989 period, and by 2020 post-1990 emissions will account for about 70 per cent of total cumulative emissions since 1850. To what extent would our contemporaries have had to change their activities and ways of life in the light of what remained a scientific hypothesis, supported by an impressive body of research and observations of different kinds, related to a major global threat? In other words, what is the liability for risk that they should now assume in the form of lower endowments of rights for the future?

Some authors, such as Gardiner (2011), have mobilized the precautionary principle to make it a moral obligation to act. But in so doing, they have given this principle a catastrophic turn, centring it on the obligation to mobilize all means to avoid the worst-case scenario. Taken up by the experts and activists who drafted the 2015 Oslo Principles (see above), this interpretation is not generally defensible, as Godard (1997) and Sunstein (2005) have shown. Indeed, there are risks behind each option for action and, in a situation of scientific uncertainty, there is no non-arbitrary means of capping the possible worst case, as demonstrated by Martin Weitzman (2009) in the case of climate change. Calibrating action according to an injunction to prevent at all cost a worst-case scenario sketched out by experts generally leads to considering extravagant solutions that are unable to attract the support of billions of people (Godard, 2015, Chapter 9).

The institutional and political translation that was traditionally given in the 1990s to the idea of prudent action regarding the climatic threat has taken the form of the Kyoto Protocol. It might therefore be thought that the citizens of the countries that have met the objectives assigned to them within this framework have fulfilled their obligations. But what about the citizens of the developed countries that have refused to adhere to this protocol (the USA), withdrawn from it (Canada), or exceeded their quotas (Japan)? And what should be asked of the populations of the countries to which the protocol has not assigned any quantitative targets for controlling their emissions? Some authors, such as Gardiner (2011) and Shue (2014), consider this protocol to be insufficient from a justice point of view, on the grounds that it would not have prevented significant damage to the most vulnerable populations. In their eyes, it is a collective moral failure. Who is responsible for this failure? Recalling the argument by Pogge (1994), these authors consider it to be attributable to the citizens of the developed countries, who have failed to incite their leaders to go further. It would therefore be the exclusive responsibility of the citizens of these countries to bear the cost of climate policies that are far more ambitious than those implemented within the international treaty framework. This should translate into far more severe cuts in the carbon budgets allocated to these countries. However, given the data set out in Chapter 4, there is no moral justification for attributing an exclusive post-1990 responsibility to the developed countries, which account for only around 42 per cent of cumulative GHG emissions since this pivotal year.

7.5 CONCLUSION

The discussion in this chapter has made it possible to emphasize the contrasting conceptions about the possible pre-existing rights of individuals and of local populations, either to the emissions of GHGs or in relation to the essential features of the local environment. The norm of the equal right of every human being to dispose of the atmosphere has emerged as a fiction that is questionable in several respects, even from a cosmopolitanist approach, and even more clearly from an approach centred on the idea of an international society of sovereign states that each have control over their natural resources. It was justifiable to take account of the level of GHG emissions reached by each country in 1990 for the subsequent allocation of emissions rights, without making this the source of an intangible structure of future rights. It has also appeared that the Lockean Provisos conditioning the legitimate appropriation of natural resources would only make sense for agents with fixed identities, which ruled out giving them trans-historical significance. Finally, the criteria of equity and of capability have both raised issues of relevance regarding the climate problem: is the organization of international action on climate protection essentially a means of increasing the resources allocated to economic development in the least-developed countries? Is it a way to make the developed countries pay just because they are richer than others? Overall, after having excluded the dubious idea of historical responsibility for emissions before 1988, no criterion has proven to be the most relevant, *in abstracto*, in terms of addressing the climate problem, and the most morally justified in allocating a carbon budget that will be tightened in the future. In order to make a choice, it would be necessary to go further in defining a hierarchy on social values and specifying main features of the situation (see Chapter 2). The equal per capita approach nevertheless presents the most fragile case in this regard, whereas that of economic and technological needs offers a stronger base. This situation paves the way for composite formulas seeking to balance interests and approaches to justice. It invites taking another look at the same issues to see whether something more conclusive can be found. This is the goal of the next chapter.

NOTES

1. Typical of this fusion of the dual position of 'should be' and 'is already" is the content of the Oslo Principles (2015), already mentioned.
2. Several scholars, on the contrary, underline the necessity to embrace climate change, development goals, the fight against poverty and international distributive justice in one strategy. See Khor (2012), Shue (2014) and Nicholas Stern (2015).

3. Curiously, moralists generally wrongly interpret the 'polluter pays' principle (PPP) as a principle of compensation for damage by the polluter, confusing this principle with a standard liability principle. They seem unaware that when the Organisation for Economic Co-operation and Development (OECD) introduced the PPP in the 1970s, it was a principle of imputation of expenses necessary to keep the environment in a satisfactory state to the polluter; it was specified not to be a liability principle (OECD, 1975). In this OECD definition, no reference is made to damage and no compensation is required for the 'victims' of pollution (Godard, 2015, Chapter 8). We may wonder why such a misunderstanding has become so widespread.
4. On this idea of 'non-envy', see Hal Varian (1974) and the book by William Baumol (1987).
5. On the 'stand-alone' criterion, see Hervé Moulin (1996). This criterion is aimed at avoiding cross-subsidization between agents sharing a common resource.
6. Based on 2008 data and an assumed elastic relationship between emissions and income, Oxfam (2015) found that: 'the poorest half of the global population – around 3.5 billion people – are responsible for only around 10 per cent of total global emissions attributed to individual consumption . . . Around 50 per cent of these emissions meanwhile can be attributed to the richest 10 per cent of people around the world, who have average carbon footprints 11 times as high as the poorest half of the population, and 60 times as high as the poorest 10 per cent. The average footprint of the richest 1 per cent of people globally could be 175 times that of the poorest 10 per cent'.

8. Climate justice in the light of justification theory

Most people hold views and beliefs on global climate justice that they defend with passion. The issue has been highly politicized within the field of international negotiations, as seen in the discourses and proposals exchanged between state representatives, as well as within the various stakeholder networks surrounding the negotiations. The term 'climate justice' has been enlisted under activist flags and given a very specific content in the ideological battle between the developing countries and the industrialized countries.

From the thorough examination of the proposals and arguments I have made in previous chapters, it appears that many proposals are ill-founded or depend on disputable postulates. The whole climate justice picture remains open and diversified, especially in terms of the degree to which issues of international wealth redistribution should be considered when building a new climate protection regime. Despite being the argument most often put forward, the exclusive historical responsibility of the developed countries for past emissions prior to 1990 has also revealed itself to be the most dubious and controversial one: objections linked to issues of ignorance, lack of control, non-identity of beneficiaries, damage, rights and retrospective illusion, to name a few, have undermined this concept. The support it has obtained from cosmopolitanist philosophers is not really confirmed by a serious discussion of the idea, despite their efforts to present rational arguments. The postulate of the rights of future unborn generations has also been shaken, being caught up in logical impossibilities (the non-existence and non-identity problems) or implausible assumptions regarding the universality of life-transcending interests of living people that would simultaneously generate duties regarding the dead and the unborn. In most cases, discussions arrive more easily at negative conclusions, dismissing an argument or a proposal rather than supporting one firmly. What I suggest in this chapter is approaching climate justice from a different theoretical universe, that of ordinary justification in situations of social interactions. It does not refer exclusively to theories built *in abstracto* from top-down universal principles and axioms indifferent to the historical reality of the organization of human societies. It considers concrete conditions of social cooperation in specific situations and the various forms it takes depending on the type of situations. The key idea is that criteria for what is just must be appropriate to the nature of the situations in which the idea of justice is mobilized. No substantive universal norm of distributive justice is appropriate to every situation.

The purpose of what French scholars Luc Boltanski and Laurent Thévenot (1991 [2006]) and Godard (1990, 1998, 2007) call the 'theory of justification' is to consider and systematize the type of arguments and tests deemed relevant by human actors to solve situations of conflict or disagreements. Claims of injustice and diversity of purpose and values are acknowledged to cause trouble or generate reluctance regarding collective action and cooperation. Justification theory is concerned by ordinary judgements communicated to others about what is just and appropriate in a socially framed situation. It also considers the accepted and instituted social tests intended to arbitrate between opposite views and projects. It is not aimed solely at obtaining a descriptive account of arguments and judgements; it explores and systematizes the representations and principles implicitly mobilized by people too. This theory thus recognizes the imperative of justification for collective and public action, but also the essential role of certain objects qualified and standardized by conventions in social coordination. Beyond arguments, this is the function of standardized tests to resolve disputes and controversies. By acknowledging the capacity of human agents for having intentions, conscience, knowledge and judgements when managing their social interactions, it differs from the issue of the involuntary collective side-effects resulting from the composition of private choices made separately, and therefore placed 'outside of justification', such as involuntary spatial segregation.

Different legitimate social orders of worth coexist within Western democratic societies as interpretation grids and social resources for coordination, each one predominantly ordering a specific social sphere (Walzer, 1983) according to a distribution shaped by history. They constitute the matrices within which questions regarding the appropriate means of coordination, the fair distribution of social positions and suitable action are conceived and solutions found. They offer social patterns that can be used to establish specific procedures and criteria needed to resolve conflicts or to overcome uncertainty about the collective path to follow. Although this framework has been developed and applied to situations taking place within a nation state, I propose using it to throw light upon the international issue of the threat of global climate change.

8.1 AN OUTLOOK ON JUSTIFICATION THEORY

8.1.1 The Model of the Ideal Polity

The objective of justification theory is to take stock of how judgements are formed about what is appropriate in a range of socially codified situations, and to link those judgements to fundamental beliefs on ideal-types of social organization. Beyond their empirical diversity, orders of worth that are deemed legitimate respond to a common 'polity' model based on the six axioms presented in Box 8.1. This structure does not describe the real functioning of Western democracies, but the ideal structure that

BOX 8.1 THE PRINCIPLES OF THE 'POLITY' MODEL

A1: The principle of common humanity

A partition of the world is established between human beings and non-human beings. Only humans have the status of subjects and are members of the polity. They recognize each other as such (symmetrical status).

A2: The principle of differentiation

There are several social states to be distributed among the members of the polity.

A3: The principle of common dignity

Each member has an equal formal power of access to the different social states, without discrimination related to birth, gender or origin (formal equality of opportunity).

A4: The order of worth

Differences in social states are turned into a hierarchy of worth, with lower and higher states, by using scales of value on the qualities of persons and/or goods attached to social states. In return, this order commands access to valuable goods and social benefits (status, reputation, and so on).

A5: The investment formula (or sacrifice principle)

Access to the highest social states requires an investment of significant resources, representing a cost or a sacrifice, in the broad sense, for the applicants.

A6: The principle of the common good

Beyond their own personal satisfaction or happiness, those who occupy the highest states generate a 'common good' that benefits all members of the 'polity', including those who remain in the lower states.

Sources: Based on Luc Boltanski and Laurent Thévenot (1991 [2006]) and Godard (1998).

the members of these societies solicit when they seek an agreement with others in a way considered legitimate by all. In this regard, the search for legitimacy expresses a renewed critical distance from the reality of social organization in a society, as well as a means of strengthening existing institutions and patterns of social relations. This is particularly the case for the principles of 'common dignity', which refer to the theme of equal opportunity, and 'the common good', which requires the elites not only to be especially lucky or clever in their own right, but to generate some good for the whole society; failing to do so would mean that the advantages from which they benefit would lose their political legitimacy.

8.1.2 Realities Linked Together within Orders of Justification

With justification theory, attention is drawn to the links between three frequently dissociated realities:

1. The presentation of arguments and information (such as results of experiments or measures) in a discourse that contributes to justifying a proposal for the person themself (personal work of deliberation and rationalization) and in the eyes of others (work of persuasion and legitimation towards others).
2. The establishment of a specific type of relationship between the actors involved in a situation that calls for decisions to be made in relation to the use, distribution or organization of a world of things – the 'goods' for economists; the type of relationship defines the range of admissible and relevant means available to actors.
3. The definition of a relationship to the world of things, which together form the objective situation in which the problems of coordination and cooperation arise; together these things structure the situation but, at the same time, they find meaning for the actors through the qualification they receive; in addition, when well qualified and standardized, some specific things provide essential support for the practical implementation of coordination procedures and, primarily, of critical tests of worth (consider the role of the chronometer and doping control in athletics).

At the juncture of these three realities (argumentation, type of relationship between human subjects, relation to a world of qualified things) are the decisive tests of justification. Tests are standardized devices organized to arbitrate or prioritize between opposing proposals or claims made by stakeholders. Examples include exams to obtain a university degree, sporting events, the economic evaluation of infrastructure projects, court trials, voting for elections, or the technical management of security issues by insurance companies or public authorities. All of these are tests of justification designed for specific purposes in specific contexts. They are clearly not interchangeable.

Not all arguments and proposals are deemed publicly acceptable in the context of pluralist democratic societies. Some ideological universes are rebutted as illegitimate for public choices. For instance, a Nazi ideology denying the same human status to some 'infrahuman' beings and more generally despising fundamental individual liberties, or a caste system which assigns an intangible social position to individuals on account of their birth, are not legitimate in democratic societies: they do not fit axioms presented in Box 8.1. Sharing this unique structure of legitimacy, various concrete embodiments give birth to different justification orders that are unequally relevant according to the nature of situations.

The steps taken either to criticize an existing state of the social world or to share certain proposals can be made within a few alternative legitimate orders of justification; each of these is articulated around a definite principle of justification that has a universalizing vocation. Thus, subjects and things are qualified through meanings that endow them with specific features: people may be 'workers', 'consumers', 'parents', 'citizens', 'union members', 'believers', and so on, according to relevant justification orders. Their values and preferences, as well as the social norms of behaviour to which they try to conform, depend on these different social roles and the situations in which they play them.

This theoretical construction delimits the possible interplay between the reference to a specific 'common good' and the choice of a defined type of relationship between people. For example, the market type of relationship cannot serve any common good.[1] The conditions for a relevant comparison between instruments of action or coordination, or public policy instruments, are clarified: not all instruments find meaning for all orders of justification; there are affinities between a given instrument and reference to a specific 'common good'; for instance, public consultations or an incentive tax are two public policy instruments whose justification references are largely different.

8.1.3 From Justification to Coordination

Justification in action depends on the correct implementation of a set of conditions and operations to achieve the expected result: effective, accepted and justified coordination. These conditions and operations are characterized as follows:

1. Each well-constituted order of justification includes the selection and qualification of the relevant things and the people empowered to implement and manipulate them. As far as environmental issues are concerned, this qualification process represents the thorny issue *par excellence*.
2. The 'common good' specific to each justification order is the reference both for the behaviour of actors within their social roles and for the definition of an order of worth over things. It enables equivalence relations to be defined and comparisons to be made, and ensures minimum predictability of conditions and means of social coordination to mobilize, which allows human agents to develop expectations and plans.
3. The dynamics of justification involve a quest for universalization, aimed at extending its sphere of influence to the whole society. Mutually exclusive orders thus call for purifying the raw situations from elements and considerations inappropriate to the main values and the vision of the world they incorporate. Confrontations between justification orders are difficult to harmonize. They typically give rise to indirect cross-denunciations about the ways of addressing a situation.
4. In a complex society, people must be able to identify which universe of justification is mainly operating in the situations they encounter, in order to be able to act in an appropriate manner. Justification processes have nothing to do with the values or preferences attached to defined groups of agents; they are not a matter of personal values or motivations, or specific community or cultural values attached to agents.
5. The highlight of justification in action is when situations focus on tests aimed at the objectification of claims and proposals, and give an answer regarding hierarchy of worth; in the end, judgements are made, certain actions are selected, new norms are adopted, social positions are assigned, or new hierarchies of worth are established. What is expected of tests is twofold: finding a suitable response to the specific circumstances of the situation requiring a judgement or a solution; but also linking this situation to a given order of worth commonly accepted by the parties involved for this type of situation.
6. Organizations such as companies, civil services, unions and associations are interpreted as special, stabilized forms of arrangements between universes (human beings, things) belonging to different orders of justification. For an organization, the weight of each order is not the same in the arrangement, this balance being different from one organization to another. One of the important characteristics of these

> arrangements is what we call a 'compromise': the actors seek to forge an agreement by referring to a new type of justification that is still virtual, but aimed at reconciling two or more pre-existing principles of justification.

In Western societies, six basic legitimate orders of worth have been identified. Each one specifies a certain representation of nature and the environment (see Box 8.2). They are designated as follows: the inspired polity, the

BOX 8.2 NATURE VIEWED WITHIN SIX ORDERS OF WORTH

The 'inspired world' takes up ideas of nature having a transcendent value and being incommensurable with the usual economic and social values. The authenticity of natural ecosystems and landscapes is at stake. Here, a religious or quasi-religious relationship with nature supports environmental protection. Practical expressions of this order in social conflicts are marked by a propensity for radical alternatives, utopias and a rejection of compromises. Argumentation refers existing, concrete situations to a 'transcendent elsewhere', such as true wilderness, an Eden or divine creation. The focus on uniqueness and invaluable natural beings is an impediment to systematic comparisons. The general form of tests in this order is to be found in embodiment and personal commitment. It gives a high standing to prophetic characters and mystic speech.

The 'domestic-traditional order' focuses on traditions in local environmental practices and a concern to rank people and things in a stable hierarchy related to the intergenerational link. The general project of transmitting a patrimony is aimed at organizing the survival of the family-like group. The value of heritage is attributed to culturally marked objects and beings (things and animals) able to embody both the identity of a cross-generational group and valuable resources enabling holders to face surprises in life. As such, if removed from their cultural and social meaning, raw nature and wilderness have the lowest ranking.

The 'fame order' embodies the idea that the value of beings depends on their fame; an action is all the more valuable as it draws the attention of a larger number of people. Public perception through the media is what really matters. Indices of notoriety (have you ever seen Hells Canyon? Do you know about it?) are the true tests to use when choosing priorities.

The 'civic order' shapes a polity based on a concern for both a basic equality of citizens, who should have equal access to the environment and natural resources, and the superiority of the general interest and public will over private interests, and of public bodies in charge of the common good over private entities: law is taken as the expression of the general will of citizens. Within this order, access to worthiness requires citizens to abstract their individual singularities for the benefit of a deliberation oriented towards the common good of society.

The 'market order' is fuelled by the desire for goods. As a collective device, the market reveals the valuable goods that people are invited to desire. It requires people and objects to be detached from their roots, land or community, to be available for playing the game of trade and market valuation. Achieving an affluent society stimulated by the satisfaction of individual desires and preferences is the value of this order. Private property rights are the means to integrate environmental assets into the market game. Nature is valuable if it can be made profitable through trade in one way (extractive uses for commodity markets) or another (tourism).

The 'industrial order' extends to the whole society and natural resources, the type of order developed by large industry during the nineteenth and twentieth centuries. It focuses on a productive use of resources and land to satisfy objectively determined needs. Knowledge-based technical performance in production and management is sought, and is expected to be reached by giving a key role to technically competent persons (engineers, experts). A strict hierarchy of social status and roles and a clear division between conception and execution are two features of this order. This is also the reign of objectivity (plans, figures, calculations, criteria); decision-making is to be based on sound science and objective data. The future is domesticated through rational plans and investment. This order basically urges that natural systems and resources be put to a productive use: there is no greater scandal than resources lying in an unproductive state.

Sources: Based on Godard (1990, 1998, 2007).

domestic polity, the polity of fame, the civic polity, the industrial polity and the market polity.

8.1.4 Compromise

A 'compromise' is a type of relatively stabilized relationship between two or several justification orders. It is incorporated into institutions or organizational rules, or a doctrine for the management of public affairs. A compromise articulates a combination between several orders and the division of spheres of influence unified by a common project and emerging new vocabulary to qualify things and agents. This is not an arrangement in the ordinary sense, which implies suspending the quest for legitimate foundations. It exemplifies what Yves Barel (1979) called 'paradoxical strategies' or 'double strategies', which manage to ensure that entities coexist, although they are normally exclusive of one another.

The paradox here is that legitimacy proceeds from several sources, whereas these sources are normally mutually exclusive. This is possible only through a dialectical game between the constituted and the virtual, which transforms the initial field of oppositions. On the one hand, the principle that serves as a reference and holds together the elements of this new order is still a virtual project. However, it is already active as a factor of rapprochement. On the other hand, this principle is fuelled by the legitimacy conferred by the constituted orders that it combines, while avoiding their mutual neutralization by installing them in a space where they are no longer considered as themselves, but as transitional components of a future synthesis.

A compromise remains a fragile thing, which can nevertheless be consolidated gradually by the progress of the virtual order that produces it and by the stabilization of new synthesis qualifications. Large institutions are generally ambiguous, in that in their uniqueness they serve simultaneously different – if not antagonistic – interests; they are organized around social compromises. This situation lends itself to denunciations based on the revelation of the interests it contains; but such denunciations, which are often partial, are not always sufficiently general in scope to cause the destruction of the institution. One of the keys to the institutional process lies in the convergence of different interests and legitimacy around a single entity.

8.1.5 On Situations

A situation is well shaped if it can be encompassed within a given order of worth. In that case, it brings together suitable subjects and objects without distortion, according to their characterization by the order through which they are viewed. Any dispute concerning the correct place to be given or the just and fair worth to be acknowledged can be solved by accepted tests that are characteristic of this order (market prices for the market order, expert opinions for the industrial order, votes and the implementation of laws for the civic order, and so on). Here, subjects and objects are not co-present by chance, as contingent events: they fit together within an order in which they have a meaning, a place and a role. There is a definite repertoire of qualifiers that constitutes a pool of language resources available to people. The latter have the cognitive responsibility of discerning the nature of the situation in which they are solicited to act and then steering their conduct by seeking appropriate coordination with others. For example, the question 'What is your price?' is quite natural and appropriate when beginning a transaction on the market, but is inappropriate and even shocking when applying for an academic degree at a university, or for political elections within the civic order: politicians are not legitimately expected to buy votes on the market.

On the contrary, an unformed, contingent situation brings heterogeneous beings together in a fortuitous manner. This is the case of atypical accidents or crises. There are no pre-established coordination procedures to resolve differences or problems in an appropriate way. A loss of meaning and the lack of appropriate benchmarks are then typical. When these traits are exacerbated, for example during a natural disaster, we call it chaos, not only because of the horror of the events affecting victims, but also because all distinctions constitutive of different orders are abolished; subjects and objects are horrendously mixed up and exposed to destructive forces. If contingency and disorder prevail, two concrete outcomes can be obtained, which have in common that they retreat from justification: the logic of power and violence, in their various forms; or the conclusion of local arrangements, a type of provisional armistice that cannot be justified, being incapable of passing the test of justification specific to each order. Thus, in these situations, illegitimacy is maintained until a new, commonly accepted test is applied.

Finally, there are intermediate situations that may potentially be framed within a given order of worth, but which also present things and subjects suitable to other orders, or remain ambiguous. This is a source of uncertainty between the qualifications and tests proposed by different orders. Which one would be the most appropriate to do justice to the situation? The problem of negotiating a new coordination regime to address the climate change issue belongs to this type of intermediate situation.

8.2 ORDERS OF WORTH AND INTERNATIONAL CLIMATE CHANGE NEGOTIATIONS

8.2.1 An Intermediate Situation

Bringing climate change into the context of international negotiations on a new global protection regime reveals the involvement of four principal orders of worth. These are the civic, industrial, domestic and market orders. The civic order reveals its presence in systematically advancing the values of equality ('citizens of the world') and the primacy of the form of universal law supposed to apply to the totality of human beings, beyond borders. The industrial order is revealed in three features: (1) the physical and technical dimensions of direct determinants of greenhouse gas (GHG) emissions, the latter being an obvious fatal by-product of the operation of physical capital and networks for production and consumption, mainly in the energy sector, but also in transportation, agriculture and forestry; (2) the importance given to scientific knowledge and expertise; and (3) the reference to economic needs. The domestic order is seen in the way in which rights claimed by certain parties are based on old practices ('grandfathering'), which amounts to a sort of emerging tradition, and in the focus on responsibility for an intergenerational future time structured by the idea of the intergenerational transmission of a precious heritage. The market order is reflected in the determination to mobilize market mechanisms to provide flexibility and efficiency by enabling full consideration of the preferences, expectations and projects of economic agents within the global regulatory regime.

Faced with this pluralism, the key question becomes: what is the core nature of the problems to be solved by international action? Ultimately, under what order of worth do they principally or essentially fall? Are these problems civic by nature, as cosmopolitanists believe? If so, precedence should be given to the concern for equality for all humans and universal requirements. A proxy often put forward (Kverndokk, 1995; Shue, 1999), but not by all cosmopolitanists,[2] is then sharing emissions quotas among states in proportion to their population size. This proposal is discussed later in this chapter.

Is this mainly a matter of improving the functioning of what could be called the global economic production machine? If the problem is considered as an 'industrial-type' issue, needs arising from current production conditions should take precedence. It would be advisable to allocate quotas derived from a global cap to countries according to their role and their contribution to the production that is under way, in order for them to correctly fulfil their productive function as far as possible (Raymond, 2003). In so doing, each country would be placed in the best position to meet the productive needs of all other countries to which they are linked by the international division of labour. The aggregate criterion that best expresses this need is therefore an index derived from gross domestic product (GDP), the measure of a country's gross economic activity. In this index, the only components that should be considered are those involving a physical activity generating carbon emissions; dematerialized services – not including the Internet, which is currently energy and carbon intensive – and speculation on real estate markets would be disregarded, for instance. Note that the developing and emerging countries that refused to enter the system of international binding pledges implicitly endorsed this needs-centred approach when they justified their refusal by their will and right to pursue their development.

Are we not dealing here with a problem defined within a traditional domestic order? The current reference to the grandfathering criterion, ensuring that the rights earned by 'grandparents' and more generally through past practices are passed on as a legacy, suggests that this reference is not irrelevant. The atmosphere has already been used for many decades by all countries, although with different intensities. It would be not only unrealistic, but also unfair to instantaneously transform these usages into a source of punishment in the form of sharply diminished allocations in the future. The atmosphere in 1990 was not a new manna to be shared amongst all inhabitants of the planet without taking account of legitimate past uses and rights. The best models for a just sharing of access to the atmosphere's capacity to absorb GHGs might then be found in cases of agrarian reform, where changes to property rights are at stake.

Could this be first and foremost a question of promoting a major extension of market-type relationships into a new domain, as denounced by those who claim that 'the world is not a commodity'? While this proposal was supported by several authors (Tickell, 2008; Caney, 2009b) and continues to be technically feasible, no negotiating party has seriously proposed that a worldwide authority put up for auction emissions rights resulting from a global cap. The money thus collected could have been deposited in an international fund or trust dedicated to the sustainable development of the planet, ensuring financial flows for the adaptation and development of the developing countries as a priority. The fact that such a procedure has never been seriously considered in the context of negotiations shows, at the very least, that dimensions other than the market are at work. Nonetheless, for the first time at the international level, the Kyoto Protocol incorporated the possibility for states to exchange quotas derived from the assigned obligations to reduce GHG emissions, and the Paris Agreement did not close the door on such exchange.

The four orders of worth involved in the global climate problem govern alternative ways of characterizing situations. They entail substantially different distributive issues and operational criteria for defining a just distribution of rights and burdens. Each of the criteria at work involves a legitimate and appropriate concept of equity within a certain order of worth, and loses its relevance outside this order. The predominant issue has shifted. It is no longer primarily a matter of directly determining which criteria (population, GDP, emissions from a certain date, historical emissions, and so on) are to be used in the situation under consideration, but of better discerning the degree of proximity between the issue of international coordination of climate policies and different orders of worth. If the question that triggered our reasoning is 'How do we proceed to a just sharing of efforts and rights?', another question logically takes precedence: 'What is the fundamental nature of the situation?'. The answer to this question will determine the most appropriate norms of justice to apply.

Regarding the issue of international climate justice, my central argument is the following: although being the preferred landmark of the developing countries and many scholars, the civic framework that places the value of per capita equalization at the forefront is certainly not the most appropriate one for the physical, technical and political nature of the situation under examination. This said, the presentation made of the way in which each main order of justification may build a facet of the international coordination problem raised by climate change suggests that we cannot avoid considering a compromise in one form or another.

In this compromise, the industrial framework should take a central position. It is well suited to the core technical attributes of the problems to be solved: GHG emissions are not directly linked to the personal existence of human beings but to their economy, their production of goods and their transportation; actions for mitigation and adaptation are first a matter of technology and investment and need to meet goals and constraints within a broader context defined by public commitments to quantitative goals, economic circuits and economic development prospects. At the same time, other references taken from the domestic and market orders cannot be dismissed and should be combined in some way with the industrial order. The domestic order would bring both the necessity to consider legitimate past behaviours, prior to 1990, regarding the use of the atmosphere, which was equally open to all countries, and concern for the transmission of global environmental heritage to future generations. As such, the market order is not relevant to define the main input variables of the climate game (which levels of concentration, then of emissions, should be achieved at different dates in order to meet the long-term objective?), but is useful in enabling flexibility, cost-efficiency and even, indirectly, distributive justice (see below).

A new compromise must be conceived around a common good that unifies all stakeholders beyond conflicts of interests and different ethical views and visions of the world. It is clearly sustainable development for all, a goal set out in the United Nations Framework Convention on Climate Change (UNFCCC), that is the relevant common good of humankind in order to address the climate issue. From the viewpoint of justice, it is a collective right to sustainable development that should be given the pivotal role of enhancement for all the main chapters of international climate negotiations: financial transfers, technological transfers, assistance in capacity-building, exchange of experience to enhance mutual learning in order to drive a huge economic and social transition to a low-carbon economy, but also the screening of industrial and infrastructure investments and of natural resource conservation projects to ensure their environmental, social and economic sustainability.

It is impossible to be more precise beyond listing these areas, since norms of justice will depend on the kind of coordination mechanisms eventually chosen by the parties to the UNFCCC. This will be demonstrated below by considering the type of regime introduced in 1997 by the Kyoto Protocol and still supported by many experts; that is, a combination of national emissions caps and the opportunity for states to trade quotas internationally in order to obtain a single carbon price. There is no better case for showing the huge change of issues that an instrumental regime may entail for the quest for justice. Before that, I must explain why the civic order does not fit the situation of international climate negotiations and is badly placed to claim the central role in a new climate compromise.

8.2.2 The Civic Order: An Allocation Proportional to Population?

The value of the equality of citizens, which is at the heart of the civic order, is often mobilized in favour of a distribution to states based on their population size, as if this procedure would amount to an actual equal per capita allocation to every human being. However, the exclusive and immediate application of an equal allocation of GHG emissions rights per inhabitant of the planet would involve such huge international transfers of resources that, realistically, no diplomat would dream of organizing immediate action on this basis. Thus, many scholars close to the negotiation scene rally to the notion of the long-term convergence of emissions rights per person, which enables them to conserve the same criterion, but for a distant future, as a perspective.

The only factor that ostensibly presents obstacles to deploying this per capita equality criterion in practice is political and economic realism. The underlying argument is the following: that which would be truly fair and equitable would not be realistic in the short term, and that which would prove realistic cannot avoid being unfair. I contend that this current view is both wrong and misleading for two reasons: (1) there are other criteria that are at least as sound and legitimate for determining an equitable allocation when they are situated in the relevant order of worth; and (2) a general, trans-order superiority of the population criterion is far from having been established.

As a matter of fact, various types of critical arguments have been raised against this per capita criterion: for instance, Posner and Sunstein (2009) and Posner and Weisbach (2010) considered that it is certainly not the best means to redistribute welfare worldwide, and Caney did not see any reason to isolate access to the atmosphere from other goods contributing to welfare.[3] What I propose below is another line of critical examination of this criterion.

Is the 'civic' framing relevant?

The choice of the 'civic' framing raises two major issues. The first is that it imposes a cosmopolitanist approach that sees in humankind a unified polity with more than 7 billion members. This issue has been discussed at length in previous chapters. The least that can be said is that it does not fit the present political structure of humankind, based on around 200 sovereign states and numerous international arrangements and institutions. International society is endowed with governance structures, but not with a fully capable single government. As for 'citizens of the world', they exist only in a metaphoric way. What does exist is a framework for negotiation between states, surrounded by many non-governmental organizations and experts trying to influence the process in one direction or another on behalf of the interests they defend. What sort of fairness can fit such a situation? States' claims to sovereignty contradict the equal per capita criterion, since each sovereign state will want to define its own social equilibrium and values without interference by others.

The second issue is whether organizing international action to protect the planet's climate is similar to giving a country's citizens the right to vote, or recognizing fundamental rights of people as free citizens. Are GHG emissions linked to the very basis of fundamental human rights, those minimal rights that must not be tampered with regardless of the benefits of doing so? For that to hold true, there must be a very close connection between GHG emissions and the existence or the dignity of human beings. It has been shown in previous chapters that this is not the case. Neither an absolute nor a sufficientarian approach to economic and social rights would be able to make the emissions of carbon dioxide (CO_2) and other gases an essential right of human persons that, in the name of equity, should be equally allocated to each human being, and thus prevail over any other consideration (Hayward, 2007; Caney, 2009b; Miller, 2009).

If one disregards GHG emissions directly resulting from human biological existence, which are not the issue at stake, one must acknowledge that technology and economic organization are essential mediators between individual economic and social rights and GHG emissions. Emissions are directly linked to technology, mainly in the field of energy, to the extent that a major variable expected to lower emissions trajectories is technological change; a huge part of economics is about the choice of a given technology among several alternatives to cover the same basic needs. Over and above technology, GHG emissions depend on the economic, social and territorial organization of each society and on the international division of labour. Many technical and organizational solutions, each with very different emissions profiles, may provide access to food, energy and sanitation.

The fact that emitting GHGs does not belong to the fundamental liberties and rights that must be ensured equally for all people is even truer in that not only can the emissions rights at stake be detached from individual people, but the carbon quotas that could be allocated to states are in fact conceived to be detached from them, even in proposals by Agarwal and Narain (1991): proposals on the negotiation table were and are considering emissions caps and abatement obligations of states, not of individuals worldwide. Moreover, since the Rio UNFCCC in 1992 (the joint implementation provision), and more explicitly with the Kyoto Protocol, emissions caps have been designed to be tradable to a large extent.[4] Readiness to trade on the market logically requires states to be able to sell the quotas in their possession without violating the rights of their people. States would not receive emissions rights with the aim and obligation of restoring them to each individual living in their territory, as transparent representatives would have to. In this respect, the foundations of the systems envisaged for the attribution and circulation of emissions rights are at odds with those of the civic order. In the latter, citizens are not entitled to sell their civic rights, for instance their voting right in elections or their access to social security benefits.

If, despite previous arguments, one persists in choosing the civic framework as the relevant reference, it is equality of states within the international society of states that should form the appropriate background.[5] Thus, it is to states that the civic equality criterion should be applied. The same emissions endowment should therefore be allocated to every state, to Belgium as well as to China, just as these countries have the same voting rights at the UN Assembly. Technically, such a solution could be considered once emissions quotas are tradable, but it has never been proposed in debates during climate negotiations. This absence must be seen as an indication that the nature of the situation clearly does not essentially meet the archetype of a civic situation within the society of states. Finally, the issue of the impossibility of determining the equal per capita share for every person across generations in an unlimited historical time frame should be raised.

Long-term convergence of emissions rights and intergenerational equality

Is it possible, without raising moral or logical objections, to readily determine the rights of current generations as a function of the goal of long-term convergence of emissions rights for every person on Earth? Going further, is it possible to develop an allocation rule based on the idea of equal per capita carbon emissions rights across generations, which is sometimes considered as the ideal of justice (Neumayer, 2000)?

I begin by examining the possibility of an individually egalitarian sharing of emissions rights among the generations concerned up to 2100. According to demographic scenarios such as those proposed by the United Nations and used by the Intergovernmental Panel on Climate Change (IPCC), for instance in its 1995 report (Bruce et al., 1996), the world population in 2100 could be between 6 and 17 billion people, that is, an approximate ratio of 1:3. Meanwhile, the cumulative quantity of CO_2 emissions between 1991 and 2100 should not exceed 2300 or 4000 $GtCO_{2eq}$, if we constrain them to reasonable expectations that GHG atmospheric concentration will not exceed 450 and 650 ppm CO_2, respectively (Bruce et al.,

1996).[6] The combination of figures mentioned would lead to an annual right to emit from around 1.8 tCO_{2eq} to 5.9 tCO_{2eq} per person on the Earth during this 100-year period. In contrast, a stabilization of the world population at its current level, combined with a moderately demanding goal (650 ppm CO_2), would permit the allocation to every person over the next century of an annual emissions level close to the one reached by the average Chinese or European citizen in 2010. A more challenging goal for atmospheric concentration, combined with high global demographic growth, would considerably reduce the individual quota to be allocated. Who can say which is the fair share of emissions rights for living human beings?

Under an egalitarian constraint, the logical structure of the intergenerational sharing of rights would imply that the current generation's right to emit GHGs depends on the size of the future population and therefore on the next generations' choices in terms of procreation. Here, we are facing something like a new, inverse 'non-identity problem': the rights of living people vary depending on the choices made in the future by young and unborn generations.

We should adopt the idea that the procreation choices of each generation engage their moral responsibility.[7] Indeed, a generation possesses the collective power to influence the number of its offspring. Demography, like economics, is ultimately built upon individual choices that can be encouraged or discouraged by public policies without infringing on rights. Most states have explicit population policies, either to encourage expansion (for instance, tax assistance for public services for young children) or to discourage it (China's former one-child policy). Governments cannot therefore plead incompetence or irresponsibility in this field. As Jonas (2000) emphasized, the right of individuals to have offspring is a weak right, one that does not generate active obligations on the part of others to ensure its achievement, and one that could be the object of public policies.

One practical consequence of this situation is that the present generation is not able to determine *ex ante* the equal share that it could claim, since this share depends not only on its own procreation choices, but also on the free procreation choices of successive generations over the next century. The practical approach typically adopted to overcome this paradox consists in broaching the question using demographic predictions, as if one were dealing with a natural phenomenon beyond human choice and responsibility. At the same time, GHG emissions are considered to be a huge political and moral issue. Such asymmetrical treatment calls for debate on moral grounds (Cafaro, 2012).

The idea of the long-term convergence of per capita emissions rights is problematic in the same way, since while the numerator (the total level of annual emissions) could be the result of a concerted political decision echoing a precautionary strategy, the denominator is affected by the same uncertainty as in the immediately egalitarian model. This is not a matter of scientific uncertainty that could be eradicated with the advancement of knowledge. The emissions rights of each generation and the range of evolutionary trajectories compatible with the goal obviously depend on the severity of the ultimate per capita emissions target, which is related to the number of inhabitants and the procreation choices of all the generations over the time frame considered.

This situation, in which the rights of present generations are dependent on the free choices made by others in the future, clearly raises the question of defining the limits of each generation's responsibility. Would it be fair to ask present generations alone – propelled by a supposedly 'ethical prudence' such as the one Shue (2009) proposed as a final word to the controversy over justice – to bear in advance the consequences of possible choices made freely after them, by others than themselves, and to determine their own rights according to what could be called a 'worst-case demographic scenario'? In so doing, they would be displaying moral hubris, since they would choose to take full responsibility for the consequences of the worst-case decisions that could possibly be made by future humans. In fact, this claim would not respect the status of the latter as moral persons; instead it would be implicitly treating them as irresponsible.

At the very least, it would be wise to implement a shift similar to the one the Brundtland Report (World Commission on Environment and Development, 1987) made in its understanding of sustainable development: the objective is to satisfy the needs and aspirations of the present without compromising the ability to meet those of the future. Reasoning about the rights of present generations should take account of the capacities of future generations. For instance, because future generations, considered from our present, will have the capacity to ensure that the world population stabilizes at around 6 billion inhabitants without infringing on people's fundamental rights, the present generation could legitimately determine its carbon emissions share based on a scenario of this type, which would authorize a level of short-term per capita emissions significantly higher than in the 'worst-case prudential' scenario. Over time, the rights of living generations should be revised as a function of the actual demographic changes observed, generation after generation. In no case can, or should, per capita emissions be equalized across time.

This reasoning makes it possible to reframe the problem to be solved. The equality to be promoted would not be that of the per capita emissions produced in each state, but that of the *ex ante* allocation of rights and burdens corresponding to specific capacities and responsibilities to be determined. A theory of just distribution that would be consistent with the fundamentals invoked (fundamental freedoms) should then set aside the criterion of strict per capita emissions within a generation and across successive generations. Thus, the very motives that could intuitively lead to putting forward the criterion of an equal distribution of rights to use the atmosphere between all human beings would, after the exercise of reason, lead to challenging such an approach. However, such instrumental directions do not suffice to justify the adoption of the civic framework as the most appropriate for the development of a new climate protection regime.

8.3 THE UPSTREAM IMPACTS OF A REGULATORY REGIME: THE CASE OF A WORLD CARBON MARKET

Norms of justice must fit the type of coordination regime chosen as the basis of the international regulatory system. Even if one distinguishes two logically different stages of regime-building – namely: (1) rules for determining rights and sharing efforts; and (2) choices of instruments for coordination and regulation – the intellectual framing of the two stages cannot be isolated. In other words, the selection of an instrumental mode of coordination cannot be relegated to a secondary and strictly instrumental compartment, to which regime designers would have recourse once arbitration concerning a just and fair allocation of rights and burdens was completed. In intermediate situations, instrumental choices have a considerable upstream impact on the agreed definition of issues of fairness and equity.

The reasons are simple enough: (1) the operational coordination regime is a key feature of the regulated situation; (2) the choice of an operational coordination regime implies a decision, whether explicit or implicit, regarding the characterization of the situation, and thus about which order or orders of worth are finally deemed the most relevant to serve as references for judgement on the fairness of allocation and burden-sharing. In this regard, the creation of an international market for carbon quotas cannot but have upstream impacts on the definition of issues of justice, as is shown now.

8.3.1 New Upstream Issues of Justice

Under an international climate protection regime based on national emissions quotas, the first issue in the search for equitable sharing is whether allocation to states is defined as a final intangible allocation, as would be the case with non-transferable national quotas, or as an initial allocation designed to be open to further exchanges of emissions rights. Except in the very unlikely event that the initial allocation is economically efficient from the outset and during the whole commitment period, the possibility of trading would lead to a reshuffling of the cards and the final distribution of actual emissions would differ greatly from the initial one.

If, by assumption, the initial allocation of quotas is made on a per capita basis, the logic of trading would be to undo this equalization. The judgement on this state of affairs depends on the norms of justice one is defending: equality of opportunity or equality of outcomes. Introducing an international carbon market can only reveal and imply that no intrinsic value is attached to the idea of an equalization of actual emissions per capita. Conversely, were such a value to exist, establishing a carbon market would be an inappropriate and morally reprehensible mode of coordination.

Next, if one agrees to see the possibility of trading as a mechanism to restore efficiency harmed by an inefficient initial allocation of rights, the very existence of a carbon market gives rise to an economic surplus resulting from the minimization of collective costs to be borne in meeting the overall GHG emissions cap. These supplementary gains will themselves pose a new problem of distributive fairness, in addition to the one raised by the initial allocation. A free market will determine how much each party will be gaining, depending on the initial marginal abatement cost level and the slope of this curve, which represents the scope of opportunities to abate emissions. If the industrialized countries face higher initial marginal abatement costs and faster-growing costs as a function of the severity of the cap than the developing countries, they will obtain the largest share of monetary gains from carbon trading.

The logic of trading in a competitive market determines respective shares in the gains from trade, even if we accept the basic idea that when it is freely agreed upon, the exchange is necessarily advantageous for the parties involved in comparison with the initial situation.[8] Indeed, in competitive conditions, the economic equilibrium of transactions in such a market is determined by both the initial rights of each party and their objective characteristics (production function, location, technological opportunities to improve carbon efficiency, and so on), synthesized by the marginal abatement cost functions.

This is what Figure 8.1 shows in the case where two groups of agents – here I am thinking of an aggregate representation of two groups of countries, North and South – have different marginal abatement cost functions, starting at point Z, which corresponds to the initial situation before any agreement on emissions abatement has been reached. Let us assume

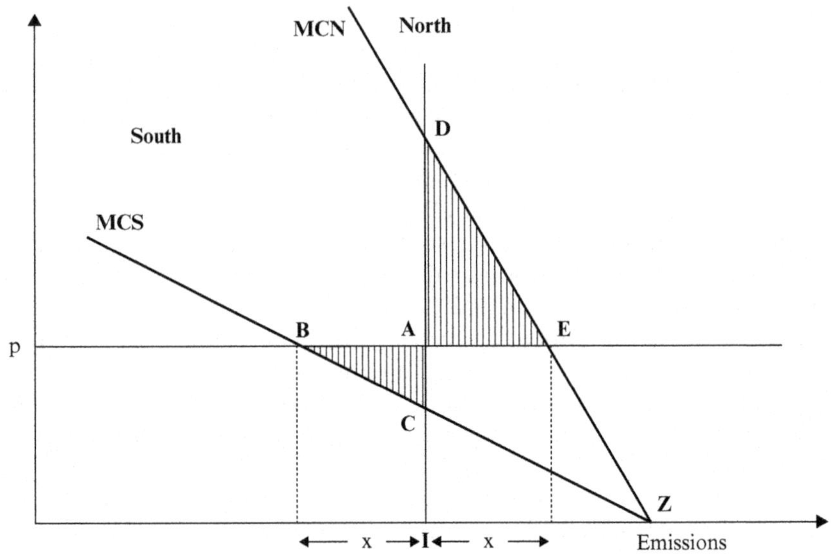

Figure 8.1 Distribution of the benefits of trade between two groups of countries

that the initial allocation of quotas from a given cap requires a parallel effort represented by point I. This allocation entails very different total costs (areas IDZ and ICZ) for the two groups of countries. We bring in a competitive market to trade emissions quotas.[9] The quantity exchanged, bought by some in the North and sold by others in the South, is *x*. In the case depicted in the figure, where the group of countries with the highest marginal abatement cost at point I, that is the buyers, is also the one for which the marginal cost function has the steepest slope on the relevant section (between points Z and I), the logic of the competitive exchange is such that the monetary benefit from trade for the buyer countries (area ADE) is higher than the gain by the seller group, whose marginal cost function has a slowly increasing slope (area ABC). This result is a geometrical affair: as one of the sides of each triangle has the same value *x*, it is the slope that determines the area of each triangle. The steeper the slope, the larger the area.

There is no reason for technological factors and market mechanisms to ensure a fair distribution of the gains from trade, whatever the consequentialist criteria we want to use; only a purely procedural conception of justice may be systematically satisfied. Let us assume an extension to the South of the regime adopted in Kyoto for the industrialized countries listed in Annex I of the Protocol. If we admit that the countries qualified as the North have both the highest marginal abatement costs at the beginning of the period and the steepest slope, then the logic of a competitive global carbon market will be to enable the North to gain more (in terms of the monetary value of avoided abatement costs) from the existence of the market than the South can obtain in terms of income gained by selling quotas. Whatever the corrections that a utility analysis may bring – $1 has more utility for poor people than for rich ones – inequality in the distribution of benefits from international carbon trading raises a specific problem of how to fairly share the benefits of trade.

Two possibilities can be envisaged to address this issue. The first is to break with the mechanisms of a competitive market by introducing, for example, the concept of 'fair trade' in emissions quotas, meaning trade with an equal share in the benefits of trade. This would imply ad hoc, administrative, non-competitive pricing that is difficult to implement; it would echo early proposals by Agarwal and Narain (1991). This solution nevertheless runs the risk of reducing the size of the cake to be shared by introducing allocative inefficiencies. The second possibility is to modify the initial allocation at the source in order to compensate in advance for the unequal sharing of benefits resulting from carbon trading. In both cases, we see that the possibility of trading, and the solution found to the problem of the equitable sharing of the benefits of trading, interfere with the content of the initial allocation considered to be just. The outcome of free trade may be deemed unfair unless it is counterbalanced by another means. If this is done *ex ante*, when proceeding to the initial allocation of quotas, by giving more quotas to parties that will gain less from trading, it will directly impair the principle of equal per capita allocation. Thus, introducing a market affects judgements on the fairness of the allocation of quotas between countries.

Other reasons concerning the political process explain why the possibility of recourse to an international carbon market modifies the behaviour of states negotiating an agreement on binding emissions limits, especially their capacity to consider criteria for redistributive justice, whatever those may be: paradoxically, it is the existence of the market that makes consideration of redistributive equity politically acceptable when states negotiate their respective rights and obligations. Without trading, each country will fight to obtain the highest level of quotas consistent with its expected needs for growth and sovereignty, with no room left for international redistributive purposes. Trading makes it acceptable to separate initial rights and needs, since it can reduce the gap between them. GHG quotas, which could be an absolute obstacle to development if not transferable, become just an additional cost to be borne by a country in order to preserve its future growth and power.

Finally, the existence of an international carbon market allows for the transformation of a physically grounded right – access to the atmospheric capacity to absorb GHGs – into access to a portion of the world's wealth, through the monetization performed by the market. This transformation itself has significant consequences for defining a truly equitable allocation of quotas. I now examine this last point.

8.3.2 A Change in the Problem of Justice

Trading turns a 'confined' problem of justice, limited to organizing access to a given physical good, into a broader problem of distributive justice whose horizon is nothing less than the fair sharing of global economic wealth. When the climate problem is considered as a confined issue whose scope can be limited to a problem of access to the atmosphere, arguments should focus on the problem as posed, without seeking to resolve the whole set of global inequalities on this occasion. This is the position vigorously defended by Posner and Weisbach (2010, 2013). An allocation aimed at satisfying current collective economic needs, combined with a concern for efficiency, may correspond to the type of legitimacy at work in the order designated as 'industrial'. For instance, each country would receive for a given period a share of the global GHG cap that would be proportional to its physically based GDP observed in the previous period or expected for the near-term future.[10] According to the relative pace of the countries' economic development, allocation would be regularly revised over time, increasing more rapidly for countries with the highest growth rate; think of China and India today. This approach would be valuable, but does not fully do justice to the situation created by the very existence of a market for carbon emissions. In fact, it would best correspond to a problem of sharing non-transferable emissions rights. In that, Posner and Weisbach (2010, 2013) support contradictory propositions, since they want to disconnect the new climate regime from global wealth redistribution goals, and at the same time to establish a world carbon market that cannot avoid the issue of the just sharing of wealth.

Establishing an international carbon market as the principal mode of coordination would turn obligations to reduce GHGs into something else, in the same way as the problem of sharing a cake amongst children is transformed when all the children have the possibility of reselling their piece of cake and of freely disposing of the money received in exchange. Trading GHG emissions quotas gives access to parts of global wealth, through the money dispensed. Consequently, what is shared through the allocation of carbon quotas is not fundamentally the physical right to emit GHGs, this right being redistributed as a function of the circumstances of the trade, but rather access to shares of global economic wealth in a generalized system of trade in goods. There is no longer any justification for focusing on the issue of equal rights of access to the atmosphere as a problem in itself.

Unfortunately, the broader problem of justice is no easier to resolve than the confined one, either in theory or in practice: we fall back on oppositions between cosmopolitanist and international statist views on justice. There is no historical indication that humanity is reaching agreement on principles for the just sharing of the entire world's wealth. At the same time, the overall market value generated by the new scarcity imposed on worldwide GHG emissions would represent only a rather small part of the total wealth created yearly. This would limit the ability to use the climate change issue as the main vehicle for organizing international transfers of wealth. Overall, the issue of global wealth redistribution cannot be dismissed, but nor can it truly be solved by the climate regime. We are typically in an intermediate situation, the references of which are unstable. It is exposed to all sorts of denunciations and the legitimacy of the norms and rules selected can only be fragile.

Thus, with an international carbon market, it would no longer be possible to limit ourselves to a confined approach to climate justice, but it is impossible to solve the new problem raised because it is dependent on the resolution of a wider and more general problem of international distributive justice. What we are now dealing with is a problem we may qualify as either 'iridescent' or 'oscillating' between confined justice and global justice. The first expression paves the way for a compromise between attributes borrowed from the two conceptual poles: the main components from a confined approach, with some others from a global one. The second expression invites us to consider the problem in dynamic terms, leading us to oscillate incessantly between the first and the second type, without a non-arbitrary stop. In other words, we must face a situation marked either by a low level of coherence or by the logical instability of the applicable standards of climate justice.

NOTES

1. For instance, Michael Sandel (2000) gained popularity with his famous slogan, 'What money can't buy'.
2. For instance, Caney (2009b) underlined the contradiction between moral arguments in favour of a direct allocation to individuals and practical aspects making the latter utopian. He then suggested a solution based on auctioning the quotas to industry and distributing the proceeds to individuals.
3. Caney (2009b) considered that equity should be sought for access to bundles of goods, rather than for each good separately. GHG emissions are just one type of good that should not be considered in isolation.

4. Meanwhile, a provision of the Kyoto Protocol limits quota transfers to those that are additional to domestic mitigation policies.
5. As previously noted, it is this shift that Rawls rightly operated when he considered the 'Law of Peoples'. His veil of ignorance then covered an assembly of state representatives, one for each state, even if further authors (Moellendorf, 2002; Brock, 2009) have jumped over the obstacle and built theories based on individual citizens of the world placed in an original position behind a veil of ignorance, and deliberating on principles of justice for humankind taken as a single global polity.
6. The acceptable level of cumulative emissions depends on the final environmental objective to be reached.
7. As Liam Murphy (2000) puts it, it is central to the values of mutual respect that we should treat other people as agents who are responsible for their own behaviour, rather than as mere forces of nature.
8. However, there could be cases where one of the participants in the trade is indifferent and the other seizes the whole benefit. Moreover, when trade generates external effects, the idea that it is still win–win for all the parties concerned could be wrong. For instance, the 'all win' assumption does not hold when one considers international differences in taxes on goods that are complementary to emissions rights. In the context of the Kyoto Protocol, this could concern differences in taxation for energy inputs. In this case, trading of emissions quotas remains advantageous for the companies that participate in the exchange, but may work to the detriment of the state in which taxation is heavier. See Godard (2001, p. 16) and Mustafa Babiker et al. (2004).
9. I do not therefore presume that each group of countries constitutes, in the exchange, a monopolistic block, which would invalidate the hypothesis of a competitive market and lead to a confrontational situation of a monopoly and a monopsony.
10. This criterion assumes that GDP is not under the direct control of climate policy in any country. Were that the case, the GDP-focused allocation rule would deliver an inappropriate incentive, since increased unsustainable growth would increase the amount of quotas received, but also the level of global GHG emissions and presumably other negative externalities.

9. Beyond justice

9.1 THE POLITICAL DEAD-END OF A CERTAIN MORALIST APPROACH

To gain a clear view of the debates and controversies on global climate justice, it was necessary to revisit the very foundations of the idea of justice. Its universe does not form a homogeneous block, but is composed of rival theories that are not articulated to the same choices or to the same primary values. Above all, these theories do not have the same conception of the place to be given to the demand for justice. Some thinkers link the idea of justice to the reign of the principle of equality. As the primary element of the constitution of a society or a polity, this principle then takes precedence over all other considerations. Other scholars make it an important requirement, but derived from the truly foundational elements, which are the conditions of social cooperation.

Constructing an abstract moral world supposed to substitute or anticipate a unified political humanity, the former see the international society of states as a contingent, arbitrary fact, without consequence for the values, arguments and standards of justice applicable. Thus, when they consider reality, they see in states pure instruments of moral requirements they believe they have identified. Valuing the moral intuition sought by parables supposed to evoke everyday life, they claim to draw valid conclusions for settling the geopolitical conflicts and disagreements raised by the global climate problem. Attentive to the unique nature of the international coordination situation, with a view to the conditions of joint action by virtually all the states of this planet, the latter consider that global climate justice cannot be deduced from the ideal norms elaborated for inter-individual relations or for the domestic order of a democratic society. This second approach demands that global justice be built for itself, taking into consideration the absence of world government and the equality of sovereign states as the central elements of the situation in which every state is invited to reason.

By positing that the contribution of all states to climate protection should be determined based on equity, their common but differentiated responsibilities and their respective capabilities, the United Nations

Framework Convention on Climate Change (UNFCCC) gave priority to the acknowledgement of differences between states and to the redistributive dimension, at the expense of conceptions of justice centred on the search for equitable rules between equals. This orientation testifies to the influence gained by a political policy of protest by the countries of the South: common action on the climate problem has been overdetermined by the North–South dispute as it existed in the 1980s and 1990s.

This initial preference enabled a set of moral fictions to expand, such as the postulate of an original common property of the atmosphere that would have founded an equal natural right of every human being to emit greenhouse gases (GHGs) since the beginning of time, and more particularly since the beginning of the industrial era. The in-depth discussion of the idea of 'historical responsibility', which is mainly attributed to the developed countries, has shown the extent of the moral and logical obstacles to its deployment and completion, despite the apparent evidence and calculations put forward by its advocates, who demonstrate an unquestionable talent for illusion. From both an empirical and a moral point of view, responsibility for the climate threat corresponds to relatively recent phenomena; since 1988 if a specific date is to be given. This responsibility is shared almost equally by the developed countries and the emerging and developing countries, if we adopt this division into two groups of countries set out in the UNFCCC and the Kyoto Protocol and attenuated, but not dismissed, by the Paris Agreement. Nothing prevents the developed countries from assuming broader political responsibility, but they are not morally bound by an exclusive historical responsibility.

Populations around the world will be affected by climate change, but the populations of the least-developed countries are those that have contributed the least to the causal emergence of the climate threat and, in most cases, are also the most vulnerable. The international society is clearly liable to those countries. Duties of assistance, demands for restorative justice for harm and obstacles generated by emissions produced since 1988, and the priority concern to cover the basic needs of every human being, converge here without difficulty and, if taken seriously, should be sufficient to awaken a sense of international political responsibility in all governments and international institutions.

As for domestic state responsibilities, emissions control policies are needed in all countries where the level of emissions is significant, if both scientific diagnoses and declarations on the objectives to be achieved in terms of the energy transition are taken seriously. In fact, what is required is solutions enabling the international society to achieve a near zero level of net carbon dioxide (CO_2) emissions by 2050.

If the international negotiating process fails to find the path to convergence on a comprehensive formal approach to global climate justice, then perhaps states should first rely on improving the procedural fairness of negotiations to ensure greater inclusiveness. In this regard, the Conference of the Parties (COP) 21 in Paris was clearly a step forward from COP 15 in Copenhagen. States should then substitute acceptability of common rules for controversial claims for justice. The huge difference between the two is linked to the reference situation (a no-agreement 'business as usual', for acceptability) and to the fact that with acceptability, what is sought is an improvement rather than a perfect ideal. The common goal should be to address the climate threat without inducing a regression of social justice, in other words generating new inequalities or increasing existing ones. If the redistributive dimension obtains some substance in the new climate regime, it will also work towards an improvement in international social justice, which should be central to the efforts of all concerned (Sen, 2009). The solution should then be sought at the intersection of the climate issue, the satisfaction of basic needs and the dynamics of economic sustainable development. But is it reasonable to expect that the parties to the climate negotiations will be able to end the arm-wrestling that has been going on for more than 20 years around the reconfiguration of geopolitical power on the international scene, and will agree to truly cooperate in order to respond to a common threat?

9.2 ALTERNATIVE PROMISES TO FUTURE GENERATIONS

The concept of justice belongs to a vast domain, that of public justification of the social states attributed to individuals and groups, and of actions committing a polity to a common destiny. In this domain, which hosts a plurality of legitimate common goods and principles for ordering things and people, situations and regulatory instruments do matter because they give orientations to the appropriate world or sphere within which normative statements are formulated on behalf of a common good.

The threat of climate change expands its spectrum to a time frame extending over decades and centuries. Such time horizons are outside the usual horizons of human action. This challenge is new and is still unresolved. Before imagining hazardous conceptual revolutions of moral theories with fragile foundations and light support among people, let us examine the resources offered by the scope of justification orders already in place. Is it not possible to find an existing framework that could seriously help to address the climate change issue?

Among the justification orders, only two incorporate an explicit approach to time. The first is the 'industrial polity', framed by values of scientific mastering of the world prolonged by technical progress and economic efficiency. This order promises safety and affluence based on the expansion of production capacities through investment and innovation. The second order is the 'domestic polity', which is organized around social relations for the biological, symbolic and social reproduction of humans over time. At the core of this polity is the reception and transmission of values and resources assumed to characterize a specific group viewed as a family, in which the biological, material and social dimensions merge. This polity is guided by values of continuity shaped by tradition and the issue of handing down resources and goods considered as patrimony.

Other polities do not offer an explicit concept of diachronic dynamics and relations. The 'market polity' calls for the permanent flexibility of agents and goods available for new transactions between co-existent people, and implies a permanent readjustment of market prices, without much living memory and with no consideration for the distant future. The 'civic polity' focuses on the formation of the general will of free and equal citizens; its preferred tests are periodic elections of representatives and opinion polls, which generate high pressure in favour of present issues and short-term changes. According to the principle of the political freedom of all generations, laws are assumed to be reversible, which does not work in favour of long-term commitments. It is only by using implausible artifices that the civic polity can serve as a reference to consider the intergenerational link between distant generations. A typical case of implausibility is provided by scholars who assume all generations of humankind are meeting under a veil of ignorance *à la* Rawls in order to define the rules of a just and fair humanity. This fiction puts aside two major features of the intergenerational situation: generations are not co-present, and former generations produce later ones, if they so wish. The 'inspired polity' refers to 'another world', which is distinct from our present world, the here and now, but this alternative world is not a future one, it is just outside our present world. Thus, we remain with the only two orders capable of promises over time.

9.2.1 The Industrial Promise: Productive Investment, Innovation and Substitution through Technology

A well-established position among most economists is the following: with few exceptions, natural resources and environmental assets have value for humans not because of what they are, but because of the services and functions they provide for human usage. In principle, they are substitutable with the help of technology through investment in new capital goods. Consequently, exploiting non-renewable resources is not in itself the source of unsustainability. The actual cause of unsustainability is the fact that the scarcity rent extracted from these resources is not reinvested into new capital at a level that is at least equivalent. Hence, it is pointless focusing on the conservation of one resource or another, or aiming to achieve the general conservation of everything that can be labelled 'natural' – in the industrial polity, beings have worth to the extent that they are put to production in order to meet human needs – but each generation should preserve forever the human capacity to produce through a sufficient level of investment and innovation.

For this to be possible and appropriate under the umbrella of sustainable development, a conceptual innovation is required, taking the form of an extension of the concept of capital beyond the standard realm of the means of production (machines, buildings, computers, facilities) and human capital (labour, knowledge and know-how, skills) to natural capital (natural resources, ecosystems that provide ecological services and amenities to human beings). It is through investment in the three forms of capital, including natural capital, via protection policies, or ecological engineering or developing knowledge in new technologies, that present generations can promise a sustainable future to future generations.

Although it is a necessary component of the contemporary world, this type of promise meets three limitations. Firstly, the time horizon of industrial business, which drives investment and can offer a credible promise, is much shorter than the extended time horizon of global environmental change (several centuries) and corresponding indirect intergenerational relations. Secondly, the industrial promise assumes managers have the capacity to truly control all technological systems and networks, however complex they may now be. To be truly meaningful, this idea of control should be applied not only in the closed and controlled spaces of industrial production, but also in the open spaces in which their output and waste is disseminated: transport, retail shops, consumers' homes, landfills and, finally, physical environments and ecosystems. History has spoken. Perfect control does not exist in this world. There is no commitment to an industrial process that does not include uncertainty regarding regular operation over decades without any failure or environmental disruption.

Environmental problems may result either from regular and accumulating flows of pollutants such as GHGs in the atmosphere, or from technological accidents and disasters. In this regard, the statement by Jonas (1984), according to which the real long-term impacts of technologies will always exceed what a future-oriented science can anticipate, is quite plausible. For instance, during the twentieth century, chlorofluorocarbons (CFCs) were first presented as particularly stable and innocuous chemicals, but since the mid-1970s they have been seen as a critical factor making the stratospheric ozone layer shrink to such a point that the only cautious solution was to forbid these gases in the hope that the ozone layer would return to pre-CFC levels about a century later. Thus, contrary to the self-perception of the industrial promise by industrial circles, the long-term perspective for industrial development is not only of a spectacular rise in production levels, but also of the increasing dissemination of polluting substances (heavy metals, chemicals, various toxic substances) and the steady erosion of biodiversity and ecosystems worldwide. There are legitimate questions about the capacity of the industrial polity to achieve a sufficient self-transformation in order to recover a capacity to make a credible promise of sustainability to future generations.

A third and final limitation is related to the distributive impacts of the industrial promise. Major technology and investment require the concentration of considerable economic means. This is not equal access to every person. The history of industrial development shows to what extent it has been based on huge inequalities in wealth and power, not only in the old capitalist countries, but also in Russia and the contemporary emerging countries. In this regard, a society focused on the industrial polity may not be compatible with conceptions of justice based on the all-pervasive pre-eminence of the value of equality.

9.2.2 The 'Domestic Promise': Patrimony to be Passed on

The domestic polity provides the central form of patrimony to qualify certain things and resources and to give content to the main social expectations of a group. This idea of patrimony combines two functions, not without tension. The first is aimed at expressing and preserving the social and cross-temporal identity of a 'patrimonial group'; it involves rules focused on the conservation and transmission of patrimony to the next generations. The second concerns the constitution and management of a pool of resources maintained as a strategic reserve enabling adaptation to changes and to unforeseen circumstances. Combining both functions imposes a cautious usage of inherited resources. Ensuring the survival of the patrimonial group over time gives rise to an imperative of handing down patrimonial resources to the next members of the group. Patrimony is not managed in the same way as capital: as Barel (1979) said, capital is managed so that it grows, whereas patrimony is managed to be handed down to future offspring, based on the expectation that these offspring will preserve the patrimony in turn (Godard, 1990).

This intellectual and normative construct acknowledges the tension between the two functions: using patrimony for one's own economic and social survival is at the expense of bequeathing it. Unity can be preserved if people who are members of a patrimonial group demonstrate certain moral qualities: they should think of themselves and be seen by the other members as provisional custodians within a lineage, a corps, a house; in other words, within an intergenerational patrimonial group embracing the dead, living and future members. The patrimonial group is the real symbolic holder of the patrimony. There are several formulas that express this same idea regarding environmental patrimony; for instance: 'we borrow the Earth from our children'. This transfer of the main quality of a person onto a transgenerational collective person makes it possible to ensure unity between the two functions, which are then intertwined.

On one hand, the present and provisional holder of the patrimony has the obligation to survive in order to effectively maintain and pass on that patrimony. To this effect, when circumstances make it necessary, living holders have the obligation to use the resources of this patrimony for their personal survival, since their disappearance would mean the interruption of transmission and the end of the patrimonial group. This is even truer given that the identity value of the patrimony may be more closely linked to the chain of personal relationships involved in the transmission than to the objective quality of the things and resources that comprise the patrimony. On the other hand, it is through the transmission chain that the patrimonial group can ensure its survival. Preserving the patrimony for its ultimate conservation by forwarding it to the next generations is the common imperative equally shared by the living members of the patrimonial group. This common concern shared by all members enables successful deliberation and cooperation on the best strategies to follow, since individual values and projects are then relegated to a position of secondary importance in relation to the accepted superior imperative of bequeathing the patrimony. In this regard, all the members share a similar commitment beyond inequalities of resources and social hierarchies. In other words, in this moral and social universe, the goal of patrimony transmission supersedes issues of distributive justice among living members of the patrimonial group. This hierarchy was evident in medieval times with peasant rules for the transmission of land assigning the inheritance to the eldest child at the expense of other children.

Overall, here we find ourselves in a universe that is fairly near to the one explored from a moral viewpoint by Thompson (2009). It is no surprise that it is vulnerable to the same tensions and fragility. Being based on the asymmetrical relation of reproduction, patrimony contains – but may also exacerbate – the correlative tension, up to the point where the legitimacy of the primacy of its principle of transmission can collapse, revealing instead arbitrary inequalities and dominance relations between living members of the group. This threat is only alleviated through the inversion of roles in the transmission chain. When the person who has received a patrimony, along with the duty to maintain it and pass it on, takes on the bequeathing role, symmetry is indirectly restored, although it does not allow direct personal reciprocity along with the transmission chain. In this order, refusing to receive a patrimony is a source of rupture as violent as the refusal to bequeath it.

Meanwhile, maintaining this risk and accepting uncertainty about the eventual success of transmission is crucial. The absolute submission of all the members of the patrimonial group to a superior imperative would only turn these members into purely instrumental means for an external cause. This would jeopardize their very quality as free subjects or moral persons. Although it may be incorporated into institutions, the patrimonial rationale should never be imposed as an external constraint on individuals. It is essential for the legitimacy of this order to be based on the voluntary commitment of people who should be given a part of autonomy to reinterpret the patrimony inherited from the past and to decide what should be passed on to the next generations. Committing to a rationale of preserving patrimony to pass on should be the accomplishment of a desire in which personal achievement embraces the idea of contributing to the life and survival of the patrimonial group over time. Transmission to future generations is not a matter of obligations of living people, but of a desire to see a family, a group, a nation or humanity persist over time. It expresses a will of some human beings to create an indirect link with members of future generations, a way of telling them that presently living people thought of them and wanted to give them the good and precious things from life on Earth.

Thus, the patrimony construct appears particularly valuable and relevant for tackling the long-term dimension of climate change. But there are also drawbacks. It is more robust and appropriate in a world where tradition dominates, since the content of patrimony is then without ambiguity: the patrimony to pass on is framed by the inherited patrimony. Its adaptive value is highest in a world that is both slow to change and highly predictable. On the contrary, in a fast-changing world where evolution touches every dimension of economic and social life, but also the physical environment – in other words, our world – tension between the two functions of patrimony intensifies: things and resources that support identity tend to lose their adaptive value. Reference to inherited patrimony may not bring much to those who consider what patrimony should be passed on to the next generations: not all old-fashioned things and customs, not all traditional natural resources have real adaptive value for the future.

The central place given to patrimony within the domestic polity has to overcome three non-trivial difficulties: (1) uncertainty about the content of the patrimony to pass on when traditional guides are of no use; (2) the asymmetry of the reproductive relationship, which requires the creation of a transgenerational collective person, the patrimonial group, in the name of which patrimony transmission may become a common aspiration; and (3) the plurality of rival patrimonial groups at the country level or worldwide, having different projects for the same objects. Since most environmental beings likely to receive patrimony status have a public dimension and cannot be simultaneously managed by different regimes at the same time, those competing patrimonial groups should find the conditions for their coexistence. For the global commons, one unique global management regime should be defined and implemented, which may nonetheless include different types of rules for subsets or local societies. The international society should elaborate procedures to achieve this.

Management for natural resources or features of the environment that would be inspired by the project of patrimony transmission should be negotiated in broad 'patrimonial contracts', transcending the specificities of each patrimonial group. But such contracts would be exposed to the risk of a loss of motivation when the features of specific identity are dissolved in broader categories up to the very general category of humankind. This final difficulty reaches its highest point at the international level, with global conditions such as the global climate.

9.3 THE LIMITS OF JUSTICE IN INTER- AND INTRAGENERATIONAL CONTEXTS

What conclusions can be drawn from this review of promises to future generations within the industrial and domestic orders? I have previously dismissed the relevance of the concept of justice to frame imaginary relations that living people can establish with their distant descendants because of the, to my mind, unescapable link of the idea of justice to the conditions of a non-altruist social cooperation. However, the irrelevance of the concept of justice for this intergenerational issue does not entail a moral withdrawal of living people regarding the future of humankind and the Earth. Other intellectual and moral support should be given to concerns for the future. Placing the issue on the grounds of justification reveals the plurality of social forms that democratic societies use to establish a link with the distant future. Two of these orders, the industrial and the domestic, provide interesting but different properties in this regard, with none of them being able to succeed alone. A good strategy would be to combine the two forms of social promises that living people can make to their distant descendants: selective conservation and transmission on one side, research and development and investment in new technology on the other, despite the ambivalence of most technological innovations. Prudence here would call for adopting a sequential approach to the future, one that would be strongly centred on the environmental patrimony to pass on to the next generations, in time sequences of two generations ahead (50 years), in the hope that the next generations will follow suit.

The concern for the justice of an international climate policy agreement certainly cannot be removed from the theory and practice of international negotiations. However, claims for a fair and just agreement are not a factor facilitating the successful conclusion of negotiations. The contradictory views put forward on behalf of justice raise additional obstacles on an already difficult road. This situation reveals the huge gap between the proposals made based on arguments founded in cosmopolitanist approaches that disregard the political organization of humankind and the actual state of the institutional equipment of the international society. In terms of justification theory, the incapacity to find an agreement on fair rules of allocation of rights, burdens and obligations reflects the lack of well-shaped, appropriate and accepted tests of worth on the international climate negotiation scene.

This gap is not an ordinary one between real and idealized states of affairs. It questions the relevance of the theoretical background itself. Negotiations on climate strategies must meet an absolute constraint: finding a system of rules and burden-sharing that major states capable of acting on the problem – in other words, those that are now the largest GHG emitters – may wish to join voluntarily in order to cooperate. On the climate scene, cooperation is not imposed by the state of the world or by a world state, and can only result from a choice. Not to cooperate on this issue is still a credible option; certainly not the best one, but sufficiently viable for the present governments of the main emitters: their vital interests are not affected in the short term by climate change and they may consider geo-engineering as an option to open the door to unilateral action, if it is deemed necessary at some point in the future. As such, the non-cooperation option would condemn the whole climate justice debate to oblivion or sterility. The structure of international relations regarding the climate change threat is radically different from the issue of defining the fair principles of justice capable of ruling a society from which individuals cannot withdraw.

This is indeed one of the lessons from the course of climate negotiations followed from Rio de Janeiro (1992) to Copenhagen (2009) and Paris (2015), via Kyoto (1997): to be able to give due consideration to the objectives of distributive justice, a tightly coordinated regime is needed around common goals and rules. At the same time, policy instruments should provide flexibility at the implementation stage, so that possession of GHG emissions quotas does not become a primary issue of the economic survival and sovereignty of nation states, in which we can see something equivalent to the fundamental freedoms of individuals in Rawls's theory. In other words, it is a tightly integrated regime, as would be the case for an extended Kyoto-type regime combining quantified commitments and international carbon trading or, alternatively, a system based on an international tax paid by each state on its emissions into a Global Fund, which would lend itself best to the inclusion of distributive justice considerations. However, the international society has chosen the bottom-up, weakly integrated option since COP 15 in Copenhagen, due to its inability for many years to agree on the goals and the basic rules of a tight regime.

The greater the importance of market mechanisms in the climate regime, the more the climate problem is deconfined, and the more issues of climate justice are dissolved into the broader issue of the worldwide redistribution of the global wealth extracted from natural resources and ecosystem services. Conversely, the more the climate regime limits itself to a loose coordination between bottom-up climate pledges and initiatives by various states, the more climate action agendas are split into various autonomous sub-areas of partial coordination, and the more implausible it becomes for considerations of international distributive justice to play a significant part.

References

Agarwal, A. and S. Narain (1991), 'Global warming in an unequal world: a case of environmental colonialism', New Delhi: Centre for Science and Environment.

Allen, M.R., D.J. Frame, C. Huntingford et al. (2009), 'Warming caused by cumulative carbon emissions towards the trillionth tonne', *Nature*, 458, 30 April, 1163–6.

Andersen, P., J.L. Andersen and H. Frost (2010), 'ITQs in Denmark and resource rent gains', *Marine Resource Economics*, 25 (1), 11–22.

Archer, David (2008), *The Long Thaw: How Humans Are Changing the Next 100,000 Years of Earth's Climate*, Princeton, NJ: Princeton University Press.

Arneson, R.J. (2005), 'Do patriotic ties limit global justice duties', *Journal of Ethics*, 9 (1/2), 127–50.

Baatz, C. (2013), 'Responsibility for the past? Some thoughts on compensating those vulnerable to climate change in developing countries', *Ethics, Policy and Environment*, 16 (1), 94–110.

Babiker, M., J. Reilly and L. Viguier (2004), 'Is international emissions trading always beneficial?', *Energy Journal*, 25 (2), 33–56.

Baer, P., J. Harte, B. Haya et al. (2000), 'Equity and greenhouse gas responsibility, Policy Forum: climate change', *Science*, 289, 29 September, 2287.

Ballantyne A.P., C.B. Alden, J.B. Miller et al. (2012), 'Increase in observed net carbon dioxide uptake by land and oceans during the past 50 years', *Nature*, 488, 2 August, 70–72.

Barel, Yves (1979), *Le Paradoxe et le système*, Grenoble: PUG.

Barrett, Scott (1991), 'Economic analysis of international environmental agreements: lessons for a global warming treaty', in OECD (ed.), *Climate Change: Selected Economic Topics*, Paris: OECD, pp. 109–49.

Barrett, Scott (2003), *Environment and Statecraft: The Strategy of Environmental Treaty-Making*, Oxford: Oxford University Press.

Barrett, S. (2008a), 'Dikes vs. windmills: climate treatise and adaptation', Discussion Paper, School of Advanced International Studies, Johns Hopkins University.

Barrett, S. (2008b), 'The incredible economics of geoengineering', *Environmental and Resource Economics*, 39, 45–54.

Barry, Brian (1978), 'Circumstances of justice and future generations', in R.I. Sikora and Brian Barry (eds), *Obligations to Future Generations*, Philadelphia, PA: Temple University Press, pp. 204–48.

Baumol, William J. (1987), *Superfairness*, Cambridge, MA: MIT Press.

Beckerman, Wilfred (2006), 'The impossibility of a theory of intergenerational justice', in Joerg Chet Tremmel (ed.), *Handbook of Intergenerational Justice*, Cheltenham and Northampton, MA: Edward Elgar Publishing, pp. 53–71.

Beckerman, Wilfred and Joanna Pasek (2001), *Justice, Posterity and the Environment*, Oxford: Oxford University Press.

Beitz, Charles R. (1979), *Political Theory and International Relations*, Princeton, NJ: Princeton University Press.

Beitz, C.R. (2005), 'Cosmopolitanism and global justice', *Journal of Ethics*, 9 (1/2), 11–27.

Bell, D. (2011), 'Global climate justice, historic emissions and excusable ignorance', *Monist*, 94 (3), 391–411.

Birnbacher, Dieter (1994), *La responsabilité envers les générations futures*, Paris: PUF.

Blomfield, M. (2013), 'Global common resources and the just distribution of emission shares', *Journal of Political Philosophy*, 21 (3), 283–304.

Bohm, P. and B. Larsen (1994), 'Fairness in a tradeable-permit treaty for carbon emissions reductions in Europe and the former Soviet Union', *Environmental and Resource Economics*, 4 (3), 219–39.

Boltanski, Luc and Laurent Thévenot (1991 [2006]), *De la justification – Les économies de la grandeur*, Paris: Gallimard; trans. (2006), *On Justification: Economies of Worth*, Princeton, NJ: Princeton University Press.

Bourg, Dominique (2006), 'The French Constitutional Charter for the environment: an effective instrument?', in Joerg Chet Tremmel (ed.), *Handbook of Intergenerational Justice*, Cheltenham and Northampon, MA: Edward Elgar Publishing, pp. 230–43.

Bovens, Luc (2011), 'A Lockean defense of grandfathering emission rights', in Denis G. Arnold (ed.), *The Ethics of Global Climate Change*, New York, NY: Cambridge University Press, pp. 124–44.

Brock, Gillian (2009), *Global Justice: A Cosmopolitan Account*, Oxford: Oxford University Press.

Brown-Weiss, Edith (1989), *In Fairness to Future Generations: International Law, Common Patrimony, and Intergenerational Equity*, Tokyo, Japan: UNU; and Dobbs Ferry, NY, USA: Transnational Publ.

Bruce, James P., Hoesung Lee and Erik F. Haites (eds) (1996), *Climate Change 1995. Economic and Social Dimensions of Climate Change, Contribution of Working Group III to the Second Assessment Report of the Intergovernmental Panel on Climate Change*, Cambridge: Cambridge University Press.

Cafaro, P. (2012), 'Climate ethics and population policy', *WIREs – Climate Change*, 3 (1), 45–61.

Caney, Simon (2005a), 'Cosmopolitanism, democracy and distributive justice', in Daniel Weinstock (ed.), *Global Justice, Global Institutions*, Calgary: University of Calgary Press, pp. 29–63.

Caney, S. (2005b), 'Cosmopolitan justice, responsibility, and global climate change', *Leiden Journal of International Law*, 18 (4), 747–75.

Caney, S. (2006), 'Environmental degradation, reparations, and the moral significance of history', *Journal of Social Philosophy*, 37 (3), 464–82.

Caney, S. (2008), 'Human rights, climate change, and discounting', *Environmental Politics*, 17 (4), 536–55.

Caney, S. (2009a), 'Climate change and the future: discounting for time, wealth, and risk', *Journal of Social Philosophy*, 40 (2), 163–86.

Caney, S. (2009b), 'Justice and the distribution of greenhouse gas emissions', *Journal of Global Ethics*, 5 (2), 125–46.

Caney, S. (2010), 'Climate change and the duties of the advantaged', *Critical Review of International Social and Political Philosophy*, 13 (1), 203–28.

Caney, Simon (2011), 'Climate change, energy rights, and equality', in Denis G. Arnold (ed.), *The Ethics of Global Climate Change*, New York, NY: Cambridge University Press, pp. 77–103.

Chakravarty, S., A. Chikkatur, H. de Coninck, S. Pacala et al. (2009), 'Sharing global CO_2 emission reductions among one billion high emitters', *PNAS*, 106 (29), 11884–8.

Chancel, L. and T. Piketty (2015), 'Carbon and inequality: from Kyoto to Paris – trends in the global inequality of carbon emissions (1998–2013) and prospects for an equitable adaptation fund', Working Paper, Paris: Iddri and Paris School of Economics, 3 November.

Chauvier, Stéphane (1999), *Justice internationale et solidarité*, Nîmes: Ed. Jacqueline Chambon.

Chauvier, Stéphane (2006), *Justice et droits à l'échelle globale*, Paris: Vrin /EHESS.

Chauvier, Stéphane (2010), 'Cosmopolitisme et acrasie collective', in Ryoa Chung and Geneviève Nootens (eds), *Le Cosmopolitisme: enjeux et débats contemporains*, Montréal: Presses Universitaires de Montréal, pp. 37–58.

Chauvier, Stéphane (2012), 'Global justice and local democracy', paper presented at the Workshop Global Justice and Democracy, Sylvie Loriaux (ed.), at Radbout University, Nijmegen, Netherlands, 27–28 September.

Chauvier, Stéphane (2013), *Éthique sans visage – Le problème des effets externes*, Paris: Vrin.

Chichilnisky, G. and G. Heal (1994), 'Who should abate carbon emissions? An international viewpoint', *Economic Letters*, 44, 443–9.

Chichilnisky, Graciela and Geoffrey Heal (eds) (2000), *Environmental Markets: Equity and Efficiency*, New York, NY: Columbia University Press.

Cline, William R. (1992), *The Economics of Global Warming*, Washington, DC: Institute for International Economics.

Coase, R. (1960), 'The problem of social cost', *Journal of Law and Economics*, 3 (October), 1–44.

Commissariat général au développement durable (CGDD) (2012), 'L'empreinte carbone des Français: évolution de 1990 à 2007', *Le Point Sur*, 114 (Mars), accessed 13 October 2016 at http://www.statistiques.developpement-durable.gouv.fr/fileadmin/documents/Produits_editoriaux/Publications/Le_Point_Sur/2012/LPS%20114%20Empreinte_carbone_1990-2007.pdf.

Copp, David (2005), 'International justice and the basic needs principle', in Gillian Brock and Harry Brighouse (eds), *The Political Philosophy of Cosmopolitanism*, Cambridge: Cambridge University Press, pp. 39–54.

Couture, Jocelyne (2010), 'Qu'est ce que le cosmopolitisme ?', in Ryoa Chung and Geneviève Nootens (eds), *Le Cosmopolitisme: enjeux et débats contemporains*, Montréal: Presses de l'Université de Montréal, pp. 15–35.

Cullet, P. (1995), 'Definition of an environmental right in a human rights context', *Netherlands Quarterly of Human Rights*, 13, 25–40.

Daniels, Norman (2016), 'Reflective equilibrium', in Edward N. Zaita (ed.), *The Stanford Encyclopedia of Philosophy*, Winter 2016 edition, accessed 21 September 2016 at http://plato.stanford.edu/archives/win2016/entries/reflective-equilibrium/.

Dasgupta, P. (2007), 'Commentary: the Stern Review's economics of climate change', *National Institute Economic Review*, 199 (1), 4–7.

Dasgupta, P. (2008), 'Discounting climate change', *Journal of Risk and Uncertainty*, 37, 141–69.

Davidson, M.D. (2008), 'Wrongful harm to future generations: the case of climate change', *Environmental Values*, 17, 471–88.

Dietz, Simon, Cameron J. Hepburn and Nicholas Stern (2008), '19. Economics, ethics and climate change', in Kaushik Basu and Ravi Kanbur (eds), *Arguments for a Better World: Essays in Honour of Amartya Sen*, Vol. 2: *Society, Institutions and Development*, Oxford: Oxford University Press, pp. 365–86.

Dimitrov, R.S. (2016), 'The Paris Agreement on climate change: behind closed doors', *Global Environmental Politics*, 16 (3), doi:10.1162/GLEP_a_00361.

Dworkin, Ronald (2000), *Sovereign Virtue: The Theory and Practice of Equality*, Cambridge, MA: Harvard University Press.

Elster, Jon (1992), *Local Justice: How Institutions Allocate Scarce Goods and Necessary Burdens*, New York, NY: Russell Sage Foundation.

den Elzen, M.G.J., G.J.O. Jos, N. Höhne et al. (2013), 'Countries' contributions to climate change: effect of accounting for all greenhouse gases, recent trends, basic needs and technological progress', *Climatic Change*, 121 (2), 397–412.

European Environment Agency (2013), 'Atmospheric concentration of CO_2', accessed 11 June 2015 at www.eea.europa.eu/data-and-maps/figures/atmospheric-concentration-of-co2-ppm-1.

Expert Group on Global Climate Obligations (2015), *Oslo Principles on Global Climate Change Obligations*, 30 March, accessed 14 April 2015 at http://globaljustice.macmillan.yale.edu/sites/default/files/files/OsloPrinciples.pdf.

Feinberg, J. (1986), 'Wrongful life and the counterfactual element in harming', *Philosophy and Social Policy*, 4 (1), 145–78.

Fleurbaey, M. and S. Zuber (2013), 'Climate policies deserve a negative discount rate', *Chicago Journal of International Law*, 13 (2), 565–95.

Frederick, S., G. Loewenstein and T. O'Donoghue (2002), 'Time discounting and time preference: a critical review', *Journal of Economic Literature*, 40 (2), 351–401.

Gardiner, Stephen M. (2011), *A Perfect Moral Storm: The Ethical Tragedy of Climate Change*, Oxford: Oxford University Press.

Godard, O. (1990), 'Environnement, modes de coordination et systèmes de légitimité: analyse de la catégorie de patrimoine naturel', *Revue économique*, 41 (2), 215–41.

Godard, O. (1992), 'Des marchés internationaux de droits à polluer pour le problème de l'effet de serre: de la recherche de l'efficacité aux enjeux de légitimité', *Politiques et Management Public*, 10 (2), 101–31.

Godard, O. (1995), 'L'environnement, du champ de recherche au concept – Une hiérarchie enchevêtrée dans la formation du sens', *Revue Internationale de Systémique*, 9(4), 405–28.

Godard, Olivier (1997), 'L'ambivalence de la précaution et la transformation des rapports entre science et décision', in Olivier Godard (ed.), *Le principe de précaution dans la conduite des affaires humaines*, Paris: Editions de la Maison des sciences de l'Homme and Inra-Editions, pp. 37–83.

Godard, Olivier (1998), 'Sustainable development and the process of justifying choices in a controversial universe', in Sylvie Faucheux, Martin

O'Connor and Jan van der Straaten (eds), *Sustainable Development: Concepts, Rationalities and Strategies*, Dordrecht: Kluwer, pp. 299–317.

Godard, Olivier (2001), *Domestic Transferable Permits for Environmental Management: Design and Implementation*, Paris: OECD Publishing.

Godard, Olivier (2007), 'Is sustainable development an alternative principle of justification?', in Sophie Thoyer and Benoît Martimort-Asso (eds), *Participation for Sustainability in Trade*, Aldershot: Ashgate, pp. 17–42.

Godard, O. (2011), 'Climate justice, between global and international justice: Insights from justification theory', EUI Working Papers RSCAS 2011/56, Florence: Robert Schuman Centre for Advanced Studies, European University Institute, November.

Godard, Olivier (2015), *Environnement et développement durable. Une approche méta-économique*, Louvain-La-Neuve: De Boeck.

Godard, O. (2016), 'Les politiques climatiques nationales: viser le "Facteur 4" en 2050?', *Economie et prévision*, 208/209, 105–33.

Godard, Olivier and Claude Henry (1998), 'Les instruments des politiques internationales de l'environnement: la prévention du risque climatique et les mécanismes de permis négociables', in Conseil d'analyse économique, *Fiscalité de l'environnement*, Paris: La Documentation française, pp. 83–174.

Godard, O. and J.-C. Hourcade (1990), 'Some instruments for collective management of climatic risks: an analysis focusing on the means for making a price signal appear', Intergovernmental Panel on Climate Change (IPCC) Seminar on Economic and Financial Measures, Paris, 26–27 April.

Goldemberg, J. and P.M. Guardabassi (2012), 'Climate change and "historical responsibilities"', *Ambiente e Sociedade*, 15 (1), 201–6.

Goldemberg, J. and P.M. Guardabassi (2015), 'Burden sharing in the implementation of the climate convention', *Energy Policy*, 81, 56–60.

Gollier, C. (2002), 'Discounting an uncertain future', *Journal of Public Economics*, 85, 149–66.

Gollier, C. (2013), 'The debate on discounting: reconciling positivists and ethicists', *Chicago Journal of International Law*, 13 (2), 549–64.

Gosseries, Axel (2004a), 'Historical emissions and free-riding', in Lukas H. Meyer (ed.), *Justice in Time: Responding to Historical Injustice*, Baden-Baden: Nomos, pp. 355–82.

Gosseries, Axel (2004b), *Penser la justice entre les générations: de l'affaire Perruche à la réforme des retraites*, Paris: Aubier.

Gosseries, Axel (2007), 'Cosmopolitan luck egalitarianism and the greenhouse effect', in Daniel Weinstock (ed.), *Global Justice, Global Institutions*, Calgary: Calgary University Press, pp. 279–309.

Gosseries, A. (2008), 'On future generations' future rights', *Journal of Political Philosophy*, 16 (4), 446–74.

Government of India (2009), 'The road to Copenhagen: India's position on climate change issues', press release, New Delhi, 27 February.

Grasso, M. (2012), 'Sharing the emission budget', *Political Studies*, 60 (3), 668–86.

Grubb, M. (1995), 'Seeking fair weather: ethics and the international debate on climate change', *International Affairs*, 71 (3), 463–96.

Guesnerie, Roger (2003), *Kyoto et l'économie de l'effet de serre*, Paris: La Documentation française.

Gupta, S. and P.M. Bhandari (1999), 'An effective allocation criterion for CO_2 emissions', *Energy Policy*, 27 (12), 727–36.

Gupta, Sujata, Dennis A. Tirpak, Nicholas Burger et al. (2007), 'Policies, instruments and co-operative arrangements', in Bert Metz, Ogunlade R. Davidson, Peter R. Bosch et al. (eds), *Climate Change 2007: Mitigation. Contribution of Working Group III to the Fourth Assessment Report of the Intergovernmental Panel on Climate Change*, Cambridge and New York, NY: Cambridge University Press, pp. 745–807.

Haas, Peter M. (1990), *Saving the Mediterranean: The Politics of International Environmental Cooperation*, New York, NY: Columbia University Press.

Halsnæs, Kirsten, Priyadarshi Shukla, Dilip Ahuja et al. (2007), 'Framing issues', in Bert Metz, Ogunlade R. Davidson, Peter R. Bosch et al. (eds), *Climate Change 2007: Mitigation of Climate Change. Contribution of Working Group III to the Fourth Assessment Report of the Intergovernmental Panel on Climate Change*, Cambridge and New York, NY: Cambridge University Press, pp. 117–67.

Hamilton, Clive (2013), *Earthmasters: The Dawn of the Age of Climate Engineering*, London: Yale University Press.

Hansen, J., P. Kharecha, M. Sato et al. (2013), 'Assessing "dangerous climate change": required reduction of carbon emissions to protect young people, future generations and nature', *PLOS ONE*, 8 (12), e81648, 3 December.

Hardin, G. (1968), 'The tragedy of the commons', *Science*, 162 (3859), 1243–8.

Harsanyi, J.C. (1953), 'Cardinal utility in welfare economics and in the theory of risk-taking', *Journal of Political Economy*, 61 (5), 434–5.

Harsanyi, J.C. (1955), 'Cardinal welfare, individualistic ethics and interpersonal comparisons of utility', *Journal of Political Economy*, 63 (4), 309–21.

Hayward, T. (2005), 'Thomas Pogge's global resources dividend: a critique and an alternative', *Journal of Moral Philosophy*, 2 (3), 317–32.

Hayward, T. (2007), 'Human rights versus emissions rights: climate justice and the equitable distribution of ecological space', *Ethics and International Affairs*, 21 (4), 431–50.

Held, David (2005), 'Principles of cosmopolitan order', in Gillian Brock and Harry Brighouse (eds), *The Political Philosophy of Cosmopolitanism*, Cambridge: Cambridge University Press, pp. 10–27.

Helm, Carsten (1999), 'Applying fairness criteria to the allocation of climate protection burdens: an economic perspective', in Ference L. Tóth (ed.), *Fair Weather: Equity Concern in Climate Change*, London: Earthscan, pp. 80–93.

Helm, C. and U.E. Simonis (2001), 'Distributive justice in international environmental policy: axiomatic foundation and exemplary formulation', *Environmental Values*, 10, 5–18.

Henry, C. (1974), 'Investment decisions under uncertainty: the irreversibility effect', *American Economic Review*, 64 (6), 1006–12.

Hoggan, James and Richard Littlemore (2009), *Climate Cover-up: The Crusade to Deny Global Warming*, Vancouver: Greystone Books.

Höhne, N., J.E. Penner, M.J. Prather et al. (2008), 'Summary report of the Ad Hoc Group for the Modelling and Assessment of Contributions to Climate Change (MATCH)', Utrecht: ECOFYS.

Howse, R. (2007), 'The concept of odious debt in public international law', UNCTAD Discussion Paper 185, (UNCTAD/OSG/DP/2007/4), July.

Huseby, R. (2015), 'Should the beneficiaries pay?', *Politics, Philosophy and Economics*, 14 (2), 209–25.

International Climate Justice Network (2002), *Bali Principles of Climate Justice*, 29 August, accessed 22 November 2016 at www.ejnet.org/ej/.

IPCC (2007), *Climate Change 2007: Synthesis Report*, Geneva, November.

Jonas, Hans (1984), *The Imperative of Responsibility: In Search of an Ethics for the Technological Age*, Chicago, IL: University of Chicago Press.

Jonas, Hans (2000), *Évolution et liberté*, Paris: Ed. Payot et Rivages.

Kanitkar, T., T. Jayaraman, M. D'Souza, P. Purkayastha et al. (2009), 'How much "Carbon Space" do we have? Physical constraints on India's climate policy and its implications', *Economic and Political Weekly*, 44 (41–42), 35–46.

Kanitkar, T., T. Jayaraman, M. D'Souza et al. (2010), 'Meeting equity in a finite carbon world: global carbon budgets and burden sharing in mitigation actions', background paper to the conference on Global Carbon Budgets and Equity in Climate Change, Tata Institute of Social Sciences, Mumbai, India, 28–29 June.

Keohane, R.O. and D.G. Victor (2016), 'Cooperation and discord in global climate policy', *Nature Climate Change*, 6 (June), 570–75.

Khor, M. (2012), 'The equitable sharing of atmospheric and development space: summary', paper presented at the UNFCCC AWG-LCA Workshop on Equity, Bonn, 16 May.
Kingston, E. (2014), 'Climate justice and temporally remote emissions', *Social Theory and Practice*, 40 (2), 281–303.
Kiss, Alexandre and Dinah Shelton (2000), *International Environmental Law*, 2nd edn, Ardsley, NY: Transnational Publishers.
Koivurova, Timo, Sébastien Duick and Leena Heinämäki (2013), 'Climate change and human rights', in Erkki J. Hollo, Kati Kulovesi and Michael Mehling (eds), *Climate Change and the Law*, Dordrecht: Springer, pp. 287–325.
Koopmans, T.C. (1960), 'Stationary ordinal utility and impatience', *Econometrica*, 28, 287–309.
Kverndokk, S. (1995), 'Tradeable CO_2 emission permits: initial distribution as a justice problem', *Environmental Values*, 4 (2), 129–48.
Kymlicka, Will (1990), *Contemporary Political Philosophy: An Introduction*, Oxford: Oxford University Press.
Le Treut, H. (2015), 'La complexité des enjeux climatiques', *Les Echos*, 26 January.
Locke, John (1690), *The Second Treatise of Civil Government*. Accessed 21 June 2015 at www.constitution.org/jl/2ndtreat.html.
Margalioth, Y. (2013), 'Analysis of the US case in climate change negotiations', *Chicago Journal of International Law*, 13 (2), 489–505.
Martinez-Alier, Joan (2002), *The Environmentalism of the Poor: A Study of Ecological Conflicts and Valuation*, Cheltenham and Northampton, MA: Edward Elgar Publishing.
Mary Robinson Foundation – Climate Justice (2011), 'Principles of Climate Justice', Dublin, accessed 21 June 2015 at www.mrfcj.org/pdf/Principles-of-Climate-Justice.pdf.
Matthews, H.D., S. Solomon and R. Pierrehumbert (2012), 'Cumulative carbon as a policy framework for achieving climate stabilization', *Philosophical Transactions of the Royal Society A*, 370 (1974), 4365–79.
Meyer, A. (2000), *Contraction and Convergence: The Global Solution to Climate Change*, Foxhole, UK: Green Books.
Meyer, L.H. (2004), 'Compensating wrongless historical emissions of greenhouse gases', *Ethical Perspectives*, 11, 20–35.
Meyer, Lukas H. (2008), 'Intergenerational Justice', in Edward N. Zalta (ed.), *The Stanford Encyclopedia of Philosophy*, accessed 13 June 2008 at http://plato.stanford.edu/archives/sum2003/entries/justice-intergenerational.
Meyer, L.H. (2013), 'Why historical emissions should count', *Chicago Journal of International Law*, 13 (2), 598–614.

Michaels, David (2008), *Doubt is Their Product: How Industry's Assault on Science Threatens Your Health*, New York, NY: Oxford University Press.

Miller, David (2007), *National Responsibility and Global Justice*, Oxford: Oxford University Press.

Miller, David (2009), 'Global justice and climate change: how should responsibilities be distributed?', in Grethe B. Peterson (ed.), *The Tanner Lectures on Human Values*, Vol. 28, Salt Lake City, UT: University of Utah Press, pp. 117–56.

Moellendorf, Darrel (2002), *Cosmopolitan Justice*, Boulder, CO: Westview Press.

Moellendorf, Darrel (2011), 'Common atmospheric ownership and equal emissions entitlements', in Denis G. Arnold (ed.), *The Ethics of Global Climate Change*, New York, NY: Cambridge University Press, pp. 104–23.

Moellendorf, Darrel (2014), *The Moral Challenge of Dangerous Climate Change: Values, Poverty and Policy*, New York, NY: Cambridge University Press.

Moulin, Hervé (1996), 'Stand alone and unanimity tests: a re-examination of fair division', in Francesco Farina, Frank Hahn and Stefano Vannucci (eds), *Ethics, Rationality, and Economic Behavior*, Oxford: Clarendon Press, pp. 121–42.

Müller, B. (2001), 'Varieties of distributive justice in climate change, an editorial comment', *Climatic Change*, 48, 273–88.

Müller, B., N. Höhne and C. Ellermann (2009), 'Differentiating (historic) responsibilities for climate change', *Climate Policy*, 9 (6), 593–611.

Murphy, Liam B. (2000), *Moral Demands in Nonideal Theory*, New York, NY: Oxford University Press.

Nagel, T. (2005), 'The problem of global justice', *Philosophy and Public Affairs*, 33 (2), 113–47.

Neumayer, E. (2000), 'In defence of historical accountability for greenhouse gas emissions', *Ecological Economics*, 33 (2), 185–92.

Nordhaus, W. (2007), 'A review of the Stern Review on the economics of climate change', *Journal of Economic Literature*, 45 (3), 686–702.

Nozick, Robert (1974), *Anarchy, State and Utopia*, New York, NY: Basic Books.

Nussbaum, Martha (2008), 'Patriotism and cosmopolitanism', in Thom Brooks (ed.), *The Global Justice Reader*, Oxford: Blackwell, pp. 306–14.

Nussbaum, Martha (2011), *Creating Capabilities. The Human Development Approach*, Cambridge, MA: Belknap Press of Harvard University Press.

Nussbaum, M. (2013), 'Climate change: why theories of justice matter', *Chicago Journal of International Law*, 13 (2), 469–88.

OECD (1975), *The Polluter Pays Principle: Definition, Analysis, Implementation*, Paris: OECD Publishing.
Okereke, C. (2010), 'Climate justice and the international regime', *WIREs – Climate Change*, 1, 462–74.
Oreskes, Naomi and Erik M. Conway (2010), *Merchants of Doubt: How a Handful of Scientists Obscured the Truth on Issues from Tobacco Smoke to Global Warming*, New York, NY: Bloomsbury Press.
Oxfam (2015), 'Extreme carbon inequality', 2 December, accessed 13 December 2015 at https://www.oxfam.org/sites/www.oxfam.org/files/file_attachments/mb-extreme-carbon-inequality-021215-en.pdf.
Pacala, S. and R. Socolow (2004), 'Stabilization wedges: solving the climate problem for the next 50 years with current technologies', *Science,* 305 (5686), 968–72.
Page, Edward A. (2006), *Climate Change, Justice and Future Generations*, Cheltenham and Northampton, MA: Edward Elgar Publishing.
Page, E.A. (2012), 'Give it up for climate change: a defence of the beneficiary pays principle', *International Theory*, 4 (2), 300–330.
Pan, J. (2005), 'Meeting human development goals with low emissions: an alternative to emission caps for post-Kyoto from a developing country perspective', *International Environmental Agreements: Politics, Law and Economics*, 5, 89–104.
Paredis, Erik, Gert Goeminne, Wouter Vanhove et al. (2009), *The Concept of Ecological Debt: Its Meaning and Applicability in International Policy*, Gent: Academia Press.
Parfit, Derek (1984), *Reasons and Persons*, Oxford: Clarendon Press.
Parks, B.C. and J.T. Roberts (2010), 'Climate change, social theory and justice', *Theory, Culture and Society*, 27 (2/3), 134–66.
Pearce, Fred (2010), *The Climate Files: The Battle for the Truth About Global Warming*, London: Guardianbooks.
Peters, G.P., J.C. Minx, C.L. Weber and O. Edenhofer (2011), 'Growth in emission transfers via international trade from 1990 to 2008', *PNAS*, 108 (21), 24 May, 8903–8.
Philibert, C. and J. Pershing (2001), 'Considering the options: climate targets for all countries', *Climate Policy*, 1, 211–27.
Pickering, J. and C. Barry (2012), 'On the concept of climate debt: its moral and political value', *Critical Review of International Social and Political Philosophy*, 15 (5), 667–85.
Piguet, Frédéric-Paul (2014), *Justice climatique et interdiction de nuire*, Genève: Globethics.net Theses.
Pinguelli-Rosa, L., M.S. Muylaert and C. Pires de Campos (2003), 'The Brazilian proposal and its scientific and methodological aspects', Working Paper, IISD and Climate Change Knowledge Network, Winnipeg.

Pogge, T.W. (1994), 'An egalitarian Law of Peoples', *Philosophy and Public Affairs*, 23 (3), 195–224.
Pogge, T.W. (2004), 'The incoherence between Rawls's theories of justice', *Fordham Law Review*, 72, 1739–59.
Pope Francis (2015), *Laudato si', On Care for our Common Home*, 24 May, accessed at http://w2.vatican.va/content/francesco/en/encyclicals/documents/papa-francesco_20150524_enciclica-laudato-si.html.
Portney, Paul R. and John P. Weyant (eds) (1999), *Discounting and Intergenerational Equity*, Washington, DC, USA: Resources for the Future.
Posner, E.A. and C. Sunstein (2009), 'Should greenhouse gas permits be allocated on a per capita basis?', *California Law Review*, 97 (1), 51–93.
Posner, Eric A. and David A. Weisbach (2010), *Climate Change Justice*, Princeton, NJ: Princeton University Press.
Posner, E.A. and D.A. Weisbach (2013), 'International Paretianism: a defense', *Chicago Journal of International Law*, 13 (2), 347–58.
Quiggin, John (2008), 'Stern and his critics on discounting and climate change: an editorial essay', *Climatic Change*, 89 (3), 195–205.
Rawls, John (1971 [1999]), *A Theory of Justice*, Cambridge, MA, USA: Harvard University Press.
Rawls, J. (1993a), 'The Law of Peoples', *Critical Inquiry*, 20 (Autumn), 36–68.
Rawls, John (1993b), *Political Liberalism*, New York, NY: Columbia University Press.
Raymond, Leigh (2003), *Private Rights in Public Resources*, Washington, DC: Resources for the Future.
Renaut, Alain (2013), *Un monde juste est-il possible?*, Paris: Stock.
République française (1997), *Second National Communication of France under the Climate Convention*, Paris, November, http://unfccc.int/resource/docs/natc/france2.pdf.
Rice, J. (2007), 'Ecological unequal exchange: international trade and uneven utilization of environmental space in the world system', *Social Forces*, 8 (3), 1369–92.
Ringius, L., A. Torvanger and B. Holtsmark (1998), 'Can multi-criteria rules fairly distribute climate burdens? OECD results from three burden-sharing rules', *Energy Policy*, 26 (10), 777–93.
Risse, Mathias (2012), 'Climate change and ownership of the atmosphere', in Mathias Risse, *On Global Justice*, Princeton, NJ: Princeton University Press, pp. 187–206.
Rive, N., A. Torvanger and J.S. Fuglestvedt (2006), 'Climate agreements based on responsibility for global warming: periodic updating, policy choices, and regional costs', *Global Environmental Change*, 16 (2), 182–94.

Roberts, J. Timmons and Bradley C. Parks (2007), *A Climate of Injustice: Global Inequality, North-South Politics, and Climate Policy*, Cambridge, MA: MIT Press.

Roberts, J.T. and B.C. Parks (2009), 'Ecologically unequal exchange, ecological debt, and climate justice: the history and implications of three related ideas for new social movement', *International Journal of Comparative Sociology*, 50 (3/4), 385–409.

Roemer, John E. (1998), *Equality of Opportunity*, Cambridge, MA: Harvard University Press.

Rose, A. (1990), 'Reducing conflict in global warming policy: the potential of equity as a unifying principle', *Energy Policy*, 18 (10), 927–35.

Rose, A., B. Stevens, J. Edmonds and M. Wise (1998), 'International equity and differentiation in global warming policy', *Environmental and Resource Economics*, 12 (1), 25–51.

Sandel, Michael J. (2000), 'What money can't buy: the moral limits of markets', in Grethe B. Peterson (ed.), *The Tanner Lectures on Human Values*, Vol. 21, Salt Lake City, UT: University of Utah Press, pp. 87–122.

Schaffer, Axel and Darrel Moellendorf (2014), 'Beyond discounted utilitarianism: just distribution of costs', in Rolf Funck and Werner Rothengatter (eds), *Man, Environment and Time: Economic Interactions in Four Dimensions*, Baden-Baden: Nomos, pp. 185–203.

Schelling, T.C. (1992), 'Some economics of global warming', *American Economic Review*, 82 (1), 1–14.

Schüssler, R. (2011), 'Climate justice: a question of historic responsibility?', *Journal of Global Ethics*, 7 (3), 261–78.

Schwartz, Thomas (1978), 'Obligations to posterity', in Richard I. Sikora and Brian Barry (eds), *Obligations to Future Generations*, Philadelphia, PA: Temple University Press, pp. 3–13.

Sen, A.K. (1977), 'Rational fools: a critique of the behavioural foundations of economic theory', *Philosophy and Public Affairs*, 6, 317–44.

Sen, A.K. (1979), 'Utilitarianism and welfarism', *Journal of Philosophy*, 76 (9), 463–89.

Sen, Amartya K. (1999), *Development as Freedom*, New York, NY: Alfred A. Knopf.

Sen, Amartya K. (2009), *The Idea of Justice*, London: Penguin Books.

Sen, Amartya and Bernard Williams (eds) (1982), *Utilitarianism and Beyond*, Cambridge: Cambridge University Press.

Shelton, D. (2009), 'Common concern of humanity', *Environmental Law and Policy*, 39 (2), 83–6.

Shue, H. (1999), 'Global environment and international inequality', *International Affairs*, 75 (3), 531–45.

Shue H. (2009), 'Historical responsibility: accountability for the results of actions taken', paper presented at the SBSTA Technical Briefing: Historical Responsibility, 6th Meeting of the Ad Hoc Working Group on Long Term Cooperative Action, UNFCC, Berlin, 4 June.

Shue, Henry (2011), 'Human rights, climate change, and the trillionth ton', in Denis G. Arnold (ed.), *The Ethics of Global Climate Change*, New York, NY: Cambridge University Press, pp. 292–314.

Shue, Henry (2014), *Climate Justice: Vulnerability and Protection*, Oxford: Oxford University Press.

Shukla, Priyadarshi R. (1999), 'Justice, equity and efficiency in climate change: a developing country perspective', in Ferenc L. Tóth (ed.), *Fair Weather: Equity Concerns in Climate Change*, London: Earthscan, pp. 145–59.

Siebert, Horst (1987), *Economics of the Environment: Theory and Policy*, Berlin: Springer.

Singer, Peter (2008), 'One atmosphere', in Thom Brooks (ed.), *The Global Justice Reader*, Oxford: Blackwell, pp. 667–88.

State of Bolivia (2009), 'Climate debt: the basis of a fair and effective solution to climate change', paper presented at the SBSTA Technical Briefing: Historical Responsibility, 6th Meeting of the Ad Hoc Working Group on Long Term Cooperative Action, UNFCC, Berlin, 4 June.

Steiner, Hillel (1983), 'The rights of future generations', in Douglas MacLean and Peter G. Brown (eds), *Energy and the Future*, Totowa, NJ: Rowman & Littlefield, pp. 151–65.

Stern, Nicholas (2007), *The Economics of Climate Change: The Stern Review*, Cambridge: Cambridge University Press.

Stern, Nicholas (2015), *Why Are We Waiting? The Logic, Urgency and Promise of Tackling Climate Change*, Cambridge, MA: MIT Press.

Sterner, T. and U.M. Persson (2008), 'An even sterner review: introducing relative prices into the discounting debate', *Review of Environmental Economics and Policy*, 2 (1), 61–76.

Stewart, Richard and Jonathan B. Wiener (2003), *Reconstructing Climate Policy: Beyond Kyoto*, Washington, DC: American Enterprise Institute Press.

Sunstein, Cass R. (2005), *Laws of Fear: Beyond the Precautionary Principle*, Cambridge: Cambridge University Press.

Thompson, J. (2000), 'The apology paradox', *Philosophical Quarterly*, 50 (201), 470–75.

Thompson, Janna (2009), *Intergenerational Justice: Rights and Responsibilities in an Intergenerational Polity*, New York, NY: Routledge.

Tickell, Oliver (2008), *Kyoto2: How to Manage the Global Greenhouse*, London: Zed Books.

Tremmel, Joerg Chet (ed.) (2006), *Handbook of Intergenerational Justice*, Cheltenham and Northampton, MA: Edward Elgar Publishing.

United Nations (1992), *Framework Convention on Climate Change*, accessed at http://unfccc.int/files/essential_background_publications_htmlpdf/applications/pdf/conveng.pdf.

Vanderheiden, Steve (2008), *Atmospheric Justice: A Political Theory of Climate Change*, Oxford: Oxford University Press.

Varian, H.R. (1974), 'Equity, envy, and efficiency', *Journal of Economic Theory*, 9, 63–91.

Verhoeven, Joe (2000), *Droit international public*, Bruxelles: Larcier.

Victor, David G. (2011), *Global Warming Gridlock: Creating More Effective Strategies for Protecting the Planet*, Cambridge and New York, NY: Cambridge University Press.

Visser 't Hooft, Hendrik P. (1999), *Justice to Future Generations and the Environment*, Dordrecht: Kluwer.

Wallack, Michael (2006), 'Justice between generations: the limits of procedural justice', in Joerg Chet Tremmel (ed.), *Handbook of Intergenerational Justice*, Cheltenham and Northampton, MA: Edward Elgar Publishing, pp. 86–105.

Walzer, Michael (1983), *Spheres of Justice: A Defense of Pluralism and Equality*, New York, NY: Basic Books.

Warlenius, R., G. Pierce, V. Ramasar et al. (2015a), 'Ecological debt: history, meaning and relevance for environmental justice', EJOLT Report no. 18, January.

Warlenius, R., G. Pierce and V. Ramasar (2015b), 'Reversing the arrow of arrears: the concept of "ecological debt" and its value for environmental justice', *Global Environmental Change*, 30, 21–30.

Wei, T., S. Yang, J.C. Moore et al. (2012), 'Developed and developing world responsibilities for historical climate change and CO_2 mitigation', *PNAS*, 109 (32), 12911–15.

Weitzman, M.L. (2007), 'A review of the Stern Review of the economics of climate change', *Journal of Economic Literature*, 45 (3), 703–24.

Weitzman, M.L. (2009), 'On modeling and interpreting the economics of catastrophic climate change', *Review of Economics and Statistics*, 91, 1–19.

World Commission on Environment and Development (1987), *Our Common Future: The Brundtland Report*, Oxford: Oxford University Press.

World Meteorological Organization (2014), *WMO Greenhouse Gas Bulletin*, 10 (November).

Young, H. Peyton (1994), *Equity: In Theory and Practice*, Princeton, NJ: Princeton University Press and Russell Sage Foundation.

Young, H.P. and A. Wolf (1992), 'Global warming negotiations: does fairness matter?', *Brookings Review*, 10, 46–51.
Zaccaï, Edwin, François Gemenne and Jean-Michel Decroly (eds) (2012), *Controverses climatiques, sciences et politique*, Paris: Presses de Sciences Po.

Index